PLANT STRUCTURE AND FUNCTION

CECIE STARR/RALPH TAGGART

BIOLOGY

The Unity and Diversity of Life

TENTH EDITION

LISA STARR
Biology Illustrator

THOMSON

BROOKS/COLE

Australia • Canada • Mexico • Singapore • Spain
United Kingdom • United States

BIOLOGY PUBLISHER: Jack C. Carey

EDITOR-IN-CHIEF: Michelle Julet

DEVELOPMENTAL EDITOR: Mary Arbogast, Peggy Williams

ASSISTANT EDITOR: Suzannah Alexander

EDITORIAL ASSISTANTS: Karoliina Tuovinen, Jana Davis

MEDIA PROJECT MANAGER: Pat Waldo

TECHNOLOGY PROJECT MANAGERS: Donna Kelley, Keli Amann

MARKETING MANAGER: Ann Caven

MARKETING ASSISTANT: Sandra Perin

ADVERTISING PROJECT MANAGER: Linda Yip

SENIOR PROJECT MANAGER, EDITORIAL/PRODUCTION: Teri Hyde

PRINT/MEDIA BUYER: Karen Hunt

PERMISSIONS EDITOR: Joohee Lee

PRODUCTION SERVICE: Lachina Publishing Services, Inc.; Grace Davidson

TEXT AND COVER DESIGN: Gary Head, Gary Head Design

ART EDITOR AND PHOTO RESEARCHER: Myrna Engler

ILLUSTRATORS: Lisa Starr, Gary Head

OFFICE SUPPORT: Brad Griffin, Verbal Clark

COVER IMAGE: *From Central America, one of the tropical rain forests that may disappear in your lifetime.* Kevin Schafer/Getty Images

COVER PRINTER: Phoenix Color Corp (MD)

COMPOSITOR: Preface, Inc.; Angela Harris, John Becker

FILM HOUSE: H&S Graphics; Tom Anderson

PRINTER: Quebecor/World, Versailles

For more information about our products, contact us at:
Thomson Learning Academic Resource Center
1-800-423-0563
For permission to use material from this text, contact us by:
Phone: 1-800-730-2214   **Fax:** 1-800-730-2215
Web: http://www.thomsonrights.com

ISBN 0-534-39749-2

BOOKS IN THE BROOKS/COLE BIOLOGY SERIES

Biology: The Unity and Diversity of Life, Tenth, Starr/Taggart
Engage in Biology Online for Starr/Taggart
Biology: Concepts and Applications, Fifth, Starr
Basic Concepts in Biology, Fifth, Starr
Biology, Sixth, Solomon/Berg/Martin
Human Biology, Fifth, Starr/McMillan
Current Perspectives in Biology, 1998, Cummings
Human Physiology, Sixth, Sherwood
Fundamentals of Physiology, Second, Sherwood
Human Physiology, Fourth, Rhoades/Pflanzer
Nutrition Concepts and Controversies, Ninth, Sizer/Whitney

Laboratory Manual for Biology, Second, Perry/Morton/Perry
Laboratory Manual for Human Biology, Morton/Perry/Perry
Photo Atlas for Biology, Perry/Morton

Introduction to Cell and Molecular Biology, Wolfe
Molecular and Cellular Biology, Wolfe
Biotechnology: An Introduction, Barnum

Introduction to Microbiology, Third, Ingraham/Ingraham
Microbiology: An Introduction, Batzing

Genetics: The Continuity of Life, Fairbanks
Human Heredity, Sixth, Cummings
Introduction to Biotechnology, Barnum
Sex, Evolution, and Behavior, Second, Daly/Wilson
Gene Discovery Lab, Benfey

Invertebrate Zoology, Sixth, Ruppert/Barnes
Mammalogy, Fourth, Vaughan/Ryan/Czaplewski
Biology of Fishes, Second, Bond
Vertebrate Dissection, Ninth, Homberger/Walker

Plant Biology, Rost et al.
Plant Physiology, Fourth, Salisbury/Ross
Introductory Botany, Berg

General Ecology, Second, Krohne
Essentials of Ecology, Second, Miller
Terrestrial Ecosystems, Second, Aber/Melillo
Living in the Environment, Thirteenth, Miller
Environmental Science, Tenth, Miller
Sustaining the Earth, Sixth, Miller

Oceanography: An Invitation to Marine Science, Third, Garrison
Marine Life and the Sea, Milne
Introduction to Marine Biology, Karleskint

Problem-Based Learning Activities for General Biology, Allen/Duch
The Pocket Guide to Critical Thinking, Second, Epstein

Brooks/Cole—Thomson Learning
10 Davis Drive
Belmont, CA 94002
USA

Asia
Thomson Learning
60 Albert Street, #15-01
Albert Complex
Singapore 189969

Australia
Nelson Thomson Learning
102 Dodds Street
South Melbourne, Victoria 3205
Australia

Canada
Nelson Thomson Learning
1120 Birchmount Road
Toronto, Ontario M1K 5G4
Canada

Europe/Middle East/Africa
Thomson Learning
Berkshire House
168-173 High Holborn
London WC1 V7AA
United Kingdom

CONTENTS IN BRIEF

Highlighted chapters are included in PLANT STRUCTURE AND FUNCTION.

INTRODUCTION

1   Concepts and Methods In Biology   2

I   PRINCIPLES OF CELLULAR LIFE

2   Chemical Foundations for Cells   20
3   Carbon Compounds In Cells   34
4   Cell Structure and Function   54
5   A Closer Look at Cell Membranes   80
6   Ground Rules of Metabolism   96
7   How Cells Acquire Energy   114
8   How Cells Release Stored Energy   132

II   PRINCIPLES OF INHERITANCE

9   Cell Division and Mitosis   150
10   Meiosis   162
11   Observable Patterns of Inheritance   176
12   Human Genetics   194
13   DNA Structure and Function   216
14   From DNA to Proteins   226
15   Controls Over Genes   238
16   Recombinant DNA and Genetic Engineering   252

III   PRINCIPLES OF EVOLUTION

17   Microevolution   270
18   Speciation   292
19   The Macroevolutionary Puzzle   304
20   The Origin and Evolution of Life   326

IV   EVOLUTION AND BIODIVERSITY

21   Prokaryotes and Viruses   346
22   Protistans   364
23   Plants   384
24   Fungi   404
25   Animals: The Invertebrates   414
26   Animals: The Vertebrates   444
27   Biodiversity in Perspective   474

PRINCIPLES OF ANATOMY AND PHYSIOLOGY

28   How Plants and Animals Work   488

V   PLANT STRUCTURE AND FUNCTION

29   Plant Tissues   504
30   Plant Nutrition and Transport   522
31   Plant Reproduction   536
32   Plant Growth and Development   550

VI   ANIMAL STRUCTURE AND FUNCTION

33   Animal Tissues and Organ Systems   566
34   Integration and Control: Nervous Systems   578
35   Sensory Reception   606
36   Integration and Control: Endocrine Systems   626
37   Protection, Support, and Movement   644
38   Circulation   664
39   Immunity   686
40   Respiration   706
41   Digestion and Human Nutrition   724
42   The Internal Environment   744
43   Principles of Reproduction and Development   758
44   Human Reproduction and Development   776

VII   ECOLOGY AND BEHAVIOR

45   Population Ecology   806
46   Social Interactions   826
47   Community Interactions   844
48   Ecosystems   866
49   The Biosphere   888
50   Perspective on Humans and the Biosphere   916

APPENDIX I   CLASSIFICATION SYSTEM

APPENDIX II   UNITS OF MEASURE

APPENDIX III   ANSWERS TO SELF-QUIZZES

A GLOSSARY OF BIOLOGICAL TERMS

APPLICATIONS INDEX

DETAILED CONTENTS

PRINCIPLES OF ANATOMY AND PHYSIOLOGY

28 HOW PLANTS AND ANIMALS WORK

On High-Flying Geese and Edelweiss 488

28.1 **Levels of Structural Organization** 490
From Cells to Multicelled Organisms 490
Growth Versus Development 490
Structural Organization Has a History 490
The Body's Internal Environment 491
How Do Parts Contribute to the Whole? 491

28.2 CONNECTIONS: THE NATURE OF ADAPTATION 492
Defining Adaptation 492
Salt-Tolerant Tomatoes 492
No Polar Bears in the Desert 492
Adaptation to What? 493

28.3 **Mechanisms of Homeostasis in Animals** 494
Negative Feedback 494
Positive Feedback 495

28.4 **Does the Concept of Homeostasis Apply to Plants?** 496
Walling Off Threats 496
Sand, Wind, and the Yellow Bush Lupine 497
About Rhythmic Leaf Folding 497

28.5 **Communication Among Cells, Tissues, and Organs** 498
Signal Reception, Transduction, and Response 498
Communication in the Plant Body 498
Communication in the Animal Body 499

28.6 CONNECTIONS: RECURRING CHALLENGES TO SURVIVAL 500
Constraints on Gas Exchange 500
Requirements for Internal Transport 500
Maintaining a Solute–Water Balance 501
Requirements for Integration and Control 501
On Variations in Resources and Threats 501

V PLANT STRUCTURE AND FUNCTION

29 PLANT TISSUES

Plants Versus the Volcano 504

29.1 **Overview of the Plant Body** 506
Shoots and Roots 506
Three Plant Tissue Systems 506
Where Do Plant Tissues Originate? 507

29.2 **Types of Plant Tissues** 508
Simple Tissues 508
Complex Tissues 508
Vascular Tissues 508
Dermal Tissues 509
Dicots and Monocots—Same Tissues, Different Features 509

29.3 **Primary Structure of Shoots** 510
How Do Stems and Leaves Form? 510
Internal Structure of Stems 510

29.4 **A Closer Look at Leaves** 512
Similarities and Differences Among Leaves 512
Leaf Fine Structure 512
Leaf Epidermis 513
Mesophyll—Photosynthetic Ground Tissue 513
Veins—The Leaf's Vascular Bundles 513

29.5 **Primary Structure of Roots** 514
Taproot and Fibrous Root Systems 514
Internal Structure of Roots 514
Regarding the Sidewalk-Buckling, Record-Breaking Root Systems 515

29.6 **Accumulated Secondary Growth— The Woody Plants** 516
Woody and Nonwoody Plants Compared 516
What Happens at the Vascular Cambium? 516

29.7 **A Closer Look at Wood and Bark** 518
Formation of Bark 518
Heartwood and Sapwood 518
Early Wood, Late Wood, and Tree Rings 519
Limits to Secondary Growth 519

30 PLANT NUTRITION AND TRANSPORT

Flies for Dinner 522

30.1 **Plant Nutrients and Their Availability in Soils** 524
Nutrients Required for Plant Growth 524
Properties of Soil 524
Leaching and Erosion 525

30.2 **How Do Roots Absorb Water and Mineral Ions?** 526
Absorption Routes 526
Specialized Absorptive Structures 526
Root Hairs 526
Root Nodules 527
Mycorrhizae 527

30.3 **How Is Water Transported Through Plants?** 528
Transpiration Defined 528
Cohesion–Tension Theory of Water Transport 528

30.4 **How Do Stems and Leaves Conserve Water?** 530
The Water-Conserving Cuticle 530
Controlled Water Loss at Stomata 530

30.5 **How Are Organic Compounds Distributed Through Plants?** *532*

Translocation *532*

Pressure Flow Theory *532*

---

31 PLANT REPRODUCTION

A Coevolutionary Tale 536

31.1 **Reproductive Structures of Flowering Plants** *538*

Think Sporophyte and Gametophyte *538*

Components of Flowers *538*

Where Pollen and Eggs Develop *539*

31.2 *Focus on Health: Pollen Sets Me Sneezing* 539

31.3 **A New Generation Begins** *540*

From Microspores to Pollen Grains *540*

From Megaspores to Eggs *540*

From Pollination to Fertilization *540*

31.4 **From Zygotes to Seeds and Fruits** *542*

Formation of the Embryo Sporophyte *542*

Seeds and Fruit Formation *543*

31.5 **Dispersal of Fruits and Seeds** *544*

31.6 *Focus on Science: Why So Many Flowers and So Few Fruits?* 545

31.7 **Asexual Reproduction of Flowering Plants** *546*

Asexual Reproduction in Nature *546*

Induced Propagation *546*

---

32 PLANT GROWTH AND DEVELOPMENT

Foolish Seedlings, Gorgeous Grapes 550

32.1 **Patterns of Early Growth and Development— An Overview** *552*

How Do Seeds Germinate? *552*

Genetic Programs, Environmental Cues *552*

32.2 **What the Major Plant Hormones Do** *554*

32.3 **Adjusting the Direction and Rates of Growth** *556*

What Are Tropisms? *556*

Responses to Mechanical Stress *557*

32.4 **How Do Plants Know When To Flower?** *558*

An Alarm Button Called Phytochrome *558*

Flowering—A Case of Photoperiodism *558*

32.5 **Life Cycles End, and Turn Again** *560*

Senescence *560*

Entering Dormancy *560*

Breaking Dormancy *561*

Vernalization *561*

32.6 CONNECTIONS: GROWING CROPS AND A CHEMICAL ARMS RACE *562*

Midbrain *594*

Evolution of the Forebrain *594*

Reticular Formation *595*

Protection at the Blood–Brain Barrier *595*

APPENDIX I   CLASSIFICATION SYSTEM

APPENDIX II   UNITS OF MEASURE

APPENDIX III   ANSWERS TO SELF-QUIZZES

GLOSSARY

CREDITS AND ACKNOWLEDGMENTS

SUBJECT INDEX

INDEX OF APPLICATIONS

PREFACE

Successive revisions of this book span nearly thirty years and reflect feedback from many instructors and students. This new edition retains the concept spreads and other pedagogical features that are the hallmarks of the book.

CONCEPT SPREADS   Reading and absorbing textbook assignments for multiple courses in the same timeframe can overwhelm students. We make it easier for them to read about and understand biology by focusing on one concept at a time. We list key concepts on the first page of each chapter. Then we organize text, art, and evidence in support of each concept on two facing pages, at most. As shown below, each *concept spread* starts with a numbered tab and ends with a boldfaced summary of the key points. Students can preview the on-page summary before they read the concept spread. They can read it again to check whether they understand the key points before turning to the next concept.

Concept spreads also offer teachers flexibility in assigning topics to fit their course requirements. For example, those who spend less time on photosynthesis may bypass the spreads on properties of light and the chemiosmotic theory of ATP formation. They may or may not assign the Focus essay on the global impact of photosynthesis. All spreads and essays are part of a chapter story, but some offer more depth.

We incorporate headings within concept spreads to help students keep track of the hierarchy of information. Transitions between spreads help them follow the story. So does setting aside some details in optional illustrations for motivated students.

With concept spreads, students find assigned topics fast, and they can focus on manageable amounts of information. This makes them more confident in their capacity to absorb the material. Our approach has a tangible outcome—improved test scores.

VISUALIZING CONCEPTS   We continue to develop text and art together, as an inseparable whole. Our *"read-me-first diagrams"* are a prime example of this approach. They allow visual learners to build a mental image of a concept before reading the text details about it. Figure 34.5 in the sample pages at right shows how simple descriptions walk students step by step through these preview diagrams. Many of our millions of student readers have written in to tell

us that our approach helps them far more than a reliance on "wordless" diagrams.

Many *anatomical drawings* are integrated overviews of structure and function. Students need not jump back and forth from text, to tables, then to art, and back again to visualize how an organ system is put together and what its component parts do. We also use *zoom sequences*, from macroscopic to microscopic views, to move students visually into a system or process. For example, Figures 37.19 and 37.20 start with a ballerina's biceps and move on down through levels of skeletal muscle contraction.

Icons remind students of where art fits into the story line. For instance, icons of a cell remind students where

---

Gold numbered tabs, such as this one, identify the start of each new concept in a chapter. Other tabs (brown) in the chapter identify Focus essays. Many essays enrich the basic text by addressing medical, environmental, and bioethical issues. Others offer detailed examples of experiments to demonstrate the power of critical thinking.

34.2

HOW ARE ACTION POTENTIALS TRIGGERED AND PROPAGATED?

Action potential propagation isn't hard to follow if you already know something about the gradients across the neural membrane. And so we now build on Section 34.1.

Approaching Threshold

When you weakly stimulate a neuron at its input zone, you disturb the ion balance across its membrane, but not much. Imagine putting a bit of pressure on the skin of a snoozing cat by gently tapping a toe on it. Tissues beneath the skin surface have receptor endings—input zones of sensory neurons. Patches of plasma membrane at these endings deform under pressure and let some ions flow across. The flow slightly changes the voltage difference across the membrane. In this case, pressure has produced a graded, local signal.

influx of ions, the cytoplasmic side of the membrane becomes less negative. This causes more gates to open and more sodium to enter. The ever increasing, inward flow of sodium is a case of **positive feedback**, whereby an event intensifies as a result of its own occurrence:

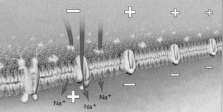

At threshold, opening of sodium gates no longer depends on the strength of the stimulus. The positive-feedback cycle is under way, and the inward-rushing sodium itself is enough to open the gated channels.

Example of a cell icon

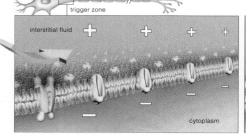

a Membrane at rest (inside negative with respect to the outside). An electrical disturbance (*yellow* arrow) spreads from an input zone to an adjacent trigger region of the membrane, which has a great number of gated sodium channels.

b A strong disturbance initiates an action potential. Sodium gates open. The sodium inflow decreases the negativity inside the neuron. The change causes more gates to open, and so on until threshold is reached and the voltage difference across the membrane reverses.

Figure 34.5   Propagation of an action potential along the axon of a motor neuron.

Graded means that signals arising at an input zone vary in magnitude. They are small to large, depending on the stimulus intensity or duration. *Local* means these signals do not spread far from the site of stimulation. Why? It takes certain kinds of ion channels to propagate a signal, and input zones simply don't have them.

When a stimulus is intense or long-lasting, graded signals spread from the input zone into an adjoining trigger zone. This patch of membrane is richly endowed with voltage-sensitive gated channels for sodium ions. *And this is where a certain amount of change in the voltage difference across the plasma membrane triggers an action potential.* The amount is the neuron's threshold level.

When these gates open, positively charged sodium ions flow into the neuron, as in Figure 34.5. With the

An All-or-Nothing Spike

Figure 34.6 shows a recording of the voltage difference across the plasma membrane before, during, and after an action potential. Notice how the membrane potential peaks once threshold is reached. All action potentials in a neuron spike to the same level above threshold as an *all-or-nothing* event. Once a positive-feedback cycle starts, nothing stops full spiking. Unless threshold is reached, the membrane disturbance subsides when the stimulation ends, and an action potential won't occur.

Each spike lasts for only a millisecond or so. Why? At the patch of membrane where the charge reversed, gated sodium channels close and shut off the sodium inflow. And about halfway into the reversal, potassium its organelles are located, as in Chapter 4. Other icons remind students of how reaction stages interconnect in a metabolic pathway, as in Chapter 7 (photosynthesis) and Chapter 8 (aerobic respiration). Others remind them of evolutionary relationships, as in Chapters 25 and 26. A multimedia icon directs students to art in the CD-ROM packaged at the back of every book. Others direct them to supplemental material on the Web and to InfoTrac® College Edition, an online database of full-length articles from 4,000 academic journals and popular sources.

Finally, we added hundreds of new *micrographs and photographs*. These are not window dressing, tacked on after the fact. They sharpen the meaning of "biodiversity" and hint at why it is worth preserving.

BALANCING CONCEPTS WITH APPLICATIONS   We draw students into each chapter with a lively or sobering application. That application gives way to a list of key concepts, an advance organizer. We attempt to maintain

their interest with focus essays, which provide depth on medical, environmental, and social issues without interrupting the conceptual flow. Where we sense that the core material needs to be livened up a bit, we weave briefer applications into the text proper. On the last pages of the book, a separate Applications Index affords fast reference to our many hundreds of applications.

FOUNDATIONS FOR CRITICAL THINKING   Like all textbooks at this level, ours helps students sharpen their capacity to think critically about nature. We walk them through experiments that yielded clear evidence in favor of or against hypotheses. The main index at the back of the book lists the experiments we selected (see the entries *Experiment, examples*, and *Test, observational*).

We also selectively use chapter introductions as well as entire chapters to show productive outcomes of critical thinking. The chapter introductions to Mendelian genetics (11), DNA structure and function (13), speciation (18), immunology (39), and behavior (46) are examples. The end of each chapter has a set of *Critical Thinking* questions. Numerous *Genetics Problems* at the end of Chapters 11 and 12 help students grasp principles of inheritance.

SUPPLEMENTS   The Instructors' Examination copy for this edition lists a comprehensive package of print and multimedia supplements, including online resources that are available to qualified adopters. Please ask your local sales representative for details.

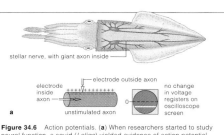

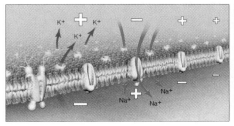

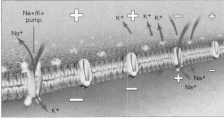

Figure 34.6 Action potentials. (**a**) When researchers started to study neural function, a squid (*Loligo*) yielded evidence of action potential spiking. This squid's "giant" axons are large enough to slip electrodes inside. (**b**) Researchers put electrodes inside and outside the axon, then stimulated it. The electrodes detected the voltage change, which showed up as deflections in a beam of light across the screen of an oscilloscope connected to the electrodes. (**c**) Typical waveform (*yellow* line) for an action potential on an oscilloscope screen.

This icon signifies that we explore the concept further on our interactive CD-ROM.

This is an example of our "read-me-first" diagrams for visual learners. They are illustrated previews of the text material, complete with simple "a b c" descriptions that guide the student each step of the way.

c With the reversal, sodium gates shut and potassium gates open (*red* arrows). Potassium follows its gradient out of the neuron. Voltage is restored. The disturbance triggers an action potential at the adjacent site, and so on, away from the point of stimulation.

d Following each action potential, the inside of the plasma membrane becomes negative once again. However, the sodium and potassium concentration gradients are not yet fully restored. Active transport at sodium–potassium pumps restores them.

channels opened, so potassium flows out. This restores the voltage difference at the patch but not the original gradients. Sodium–potassium pumps actively transport sodium back outside and potassium inside. Once this is done, most potassium gates close and sodium gates are in their original position—ready to be opened with the arrival of a suitable signal at the membrane patch.

The Direction of Propagation

During an action potential, the inward rush of sodium ions affects the charge distribution across the adjacent membrane patch, where an equivalent number of gated channels open. Gated channels open in the *next* patch, and the next, and so on. This positive feedback event is self-propagating and does not diminish in magnitude. You might be wondering: Do action potentials spread

back to the trigger zone? No. For a brief period after the inward rushing of sodium ions, the voltage-gated channels remain insensitive to stimulation, so sodium ions cannot move through them. This is one reason why action potentials do not spread back to the patch of membrane where they were initiated. It is why they propagate themselves away from it.

Ions cross the neural membrane through transport proteins that serve as gated or open channels. At a suitably disturbed trigger zone, sodium gates open in an all-or-nothing way, and the inward-rushing sodium causes an action potential. Sodium–potassium pumps restore the original ion gradients.

Sodium gates across the membrane are briefly inactivated after an action potential, which is one reason why an action potential is self-propagating away from a trigger zone.

Boldfaced key concepts end each tabbed section. Students can read them to preview the section's main points, then read them again to reinforce what they learn from the section.

All together, these highlighted statements are a running summary of the take-home lessons.

This icon reminds students to check out the website, which expands on the section's topic.

Principles of Anatomy and Physiology

Two bottlenose dolphins leaping from the sea, to which they are most exquisitely adapted, into air—where they cannot survive for long. This transient bridging of two very different environments invites you to ask: Are there fundamental constraints on how any organism is put together and how it functions, regardless of where it lives? And can we identify patterns in the diverse responses to recurring challenges?

28

HOW PLANTS AND ANIMALS WORK

On High-Flying Geese and Edelweiss

Each year, Mount Everest beckons irresistibly to some humans simply because it is the highest place on Earth (Figure 28.1). Being the highest place on Earth, it has the thinnest air. Climbers must condition themselves for months before their assault on the summit. They make short-term changes in how they breathe and how many oxygen-transporting red blood cells they produce. Even then, breathing near the summit will be agonizing. A few climbers won't make it back.

If oxygen is so scarce, then how does the barheaded goose (*Anser indicus*) regularly wing its way back and forth over the highest peaks of the Himalayas? How does its geographically separate relative, the Andean goose (*Cloephaga melanoptera*), live comfortably at 6,000 meters above sea level in the Andes of South America?

The type of hemoglobin synthesized by their red blood cells binds oxygen more strongly than human hemoglobin is able to do. Yet their hemoglobin is the same as ours, except for one amino acid substitution.

Which type of amino acid was substituted differs between the two kinds of geese. But both types disrupt the same weak interaction in two of four polypeptide chains that make up the hemoglobin molecule. The outcome, a greater oxygen-binding capacity, helps keep both birds flying high in different parts of the world.

Look about and you also find plants that survive brutal conditions at 4,100 meters (13,500 feet) above sea level, higher than the timberline of the Himalayan range. One species, *Leontopodium alpinum*, is the more ancient and equally endangered relative of Europe's edelweiss (Figure 28.2). It is anchored in pockets of soil in tiny fissures in rocks, where its roots are protected from soil contractions and expansions each time the below-freezing winter weather alternates with spring thaws. Its aboveground parts form short, tight cushions close to the rocks, and its sturdy stems resist the strong mountain winds. Its flowers appear woolly, so thickly are they covered with fine epidermal hairs. The hairs protect floral reproductive structures from ultraviolet radiation, which is intense at high altitudes. They also help the plant resist losing moisture to winds.

These examples can start you thinking about how plants and animals are put together and how their body parts function in particular environments. Because each human, barheaded goose, and alpine plant shares the same general characteristics with others of its species,

Figure 28.1 A climber near the summit of Mount Everest, where the simple act of taking in enough oxygen becomes torture. Yet the barheaded goose (Figure 28.2*a*) flies over the Himalaya range every year. It migrates back and forth between lakes, marshes, and rivers in high mountain valleys to warmer wintering grounds in India.

Figure 28.2 (**a**) Barheaded goose (*Anser indicus*). (**b**) *Leontopodium alpinum*, a low-growing plant that evolved above the timberline in the Himalayas.

you might infer that structural traits have a genetic basis and are vulnerable to mutation. You also might infer that long-standing adaptations are outcomes of natural selection and offer a better fit with prevailing environmental conditions. You might infer that the individuals of at least some species besides humans can adapt briefly to more extreme conditions than they are used to. You may also find the examples reinforce what you already suspect—*that the structure of a body part typically correlates with a present or past function.*

Such thoughts are your passport to the world of anatomy and physiology.

Anatomy is the study of an organism's form—that is, its morphology—from its molecular foundation to its physical organization as a whole. **Physiology** is the study of patterns and processes by which an organism survives and reproduces in the environment. It deals with how structures are put to use and the nature of metabolic and behavioral adjustments to changing conditions. It also deals with how, and to what extent, physiological processes can be controlled.

This chapter introduces some important concepts. Study them well, for you will apply them in the next two units of the book. They can help deepen your sense of how plants and animals function under stressful as well as favorable environmental conditions.

Key Concepts

1. Anatomy is the study of body form at many levels of structural organization, from molecules through tissues, organs and, for most animals, organ systems.

2. Physiology is the study of how the body functions in response to predictable and unexpected aspects of the environment.

3. The structure of most body parts correlates with current or past functions. Typical examples are tissues that cover the plant or animal body or line its internal parts; structures such as woody stems that function as scaffolds for upright growth; and thick, strong leg bones with load-bearing functions.

4. Most aspects of body form and function are long-term adaptations. They are heritable features that evolved in the past and have continued to be effective under prevailing conditions. Individuals of many species also adapt relatively quickly, over the short term, to stressful conditions.

5. For multicelled organisms, the extracellular fluid that bathes living cells is an internal environment. Cells, tissues, organs, and organ systems typically contribute to maintaining the internal environment within a range that individual cells can tolerate. This concept helps us understand the functions and interactions of body parts.

6. Homeostasis is the name for stable operating conditions in the internal environment. Negative and positive feedback mechanisms are among the controls that work to maintain these conditions.

7. Cells of tissues and organs communicate with one another by secreting diverse hormones and other signaling molecules into extracellular fluid, and selectively responding to signals from other cells.

LEVELS OF STRUCTURAL ORGANIZATION

From Cells to Multicelled Organisms

In most plants and animals, cells, tissues, organs, and organ systems split up the work and contribute to the survival of the body as a whole. A separate lineage of cells gives rise to the body parts that will function in reproduction. The body shows a **division of labor**.

A **tissue** is a community of cells and intercellular substances, all interacting in one or more tasks. For example, wood tissue and bony tissue both function in structural support. An **organ** consists of two or more tissues, organized in specific proportions and patterns, that perform one or more related tasks. A leaf adapted for photosynthesis, an eye that responds to light in the surroundings—these are examples of organs. An **organ system** is composed of different organs that interact physically, chemically, or both during common tasks. A plant's shoot system, with organs of photosynthesis and reproduction, is like this (Figure 28.3). And so is the digestive system of an animal, which takes in food, breaks it down, absorbs the breakdown products, and gets rid of the residues and wastes.

Growth Versus Development

The structural organization of a plant or animal emerges during growth and development. Be sure you understand at the outset that these two terms are not the same. Generally, **growth** of a multicelled organism means its cells increase in number, size, and volume. **Development** refers to the successive stages in the formation of specialized tissues, organs, and organ systems. In other words, we measure growth in *quantitative* terms, and we measure development in *qualitative* terms.

Figure 28.3 Morphology of a tomato plant (*Lycopersicon*). Different cell types make up various vascular tissues that conduct either water and dissolved minerals or organic substances. They thread through other tissues that make up most of the plant. A different tissue covers surfaces of the root and shoot systems.

Structural Organization Has a History

The structural organization of each tissue, organ, and organ system has an evolutionary history. Think back on how plants invaded the land, and you have clues to how their structures reflect specific functions. Plants that left aquatic habitats found plenty of sunlight and carbon dioxide for photosynthesis, oxygen for aerobic respiration, and mineral ions. As they dispersed farther away from their aquatic cradle, however, they faced a new challenge: how to keep from drying out in air.

Think about that when you look at micrographs of cellular pipelines in the roots, stems, and leaves of a typical plant (Figure 28.3). They conduct tiny streams of water from soil to the leaves. Stomata, small gaps across the leaf epidermis, open and close in ways that help conserve water (Section 7.7). Root systems grow

reproductive organ
(tomato flower)

stem tissues, cross-section, for support, storage, water distribution, food distrbution

root water-conducting cells, longitudinal section

SHOOT SYSTEM
(aboveground parts)

ROOT SYSTEM
(belowground parts, mostly)

Figure 28.4 Some of the structures that function in human respiration. Cells carry out specialized tasks, as when ciliated cells sweep bacteria and other airborne particles out of airways to and from two organs, a pair of lungs. Tissues arranged as tubes (blood capillaries), a fluid connective tissue called blood, and epithelial tissue in the form of thin air sacs function in gas exchange inside the lungs. Other organs, including the airways, are components of this organ system.

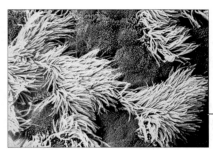

individual cells, ciliated and mucus-secreting, that line respiratory airways

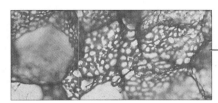

lung tissue (tiny air sacs) laced with blood capillaries—one-cell-thick tubular structures that hold blood, which is a fluid connective tissue

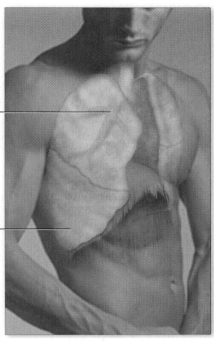

organs (lungs), part of an organ system (respiratory tract) of a whole organism

outward to places where water and minerals are more concentrated in soil. Inside the stems, cells with lignin-reinforced walls collectively support upright growth.

Similarly, respiratory systems of land animals are adaptations to life in air. Gases move into and out of the body by diffusing across a moist surface. This is not a problem for aquatic organisms. However, moist surfaces dry out in air. Land animals have moist sacs inside their body for gas exchange (Figure 28.4). And this brings us to the *internal* environment.

The Body's Internal Environment

To stay alive, plant and animal cells must be bathed in a fluid that offers nutrients and carries away metabolic wastes. In this they are no different from free-living cells. But plants and animals consist of thousands to trillions of cells. Each cell must draw nutrients from and dump wastes into the fluid bathing them.

Body fluids *not* inside cells—extracellular fluids—are an **internal environment**. Functionally, changes in the composition and volume of this environment affect cell activities. The type and number of ions are vital. They must be kept at concentrations compatible with metabolism. It makes no difference whether the plant or animal is simple or complex. *Its components require a stable fluid environment for all of its living cells.* As you will see, this concept is absolutely central to learning how plants and animals work.

How Do Parts Contribute to the Whole?

The next two units describe how each plant or animal carries out these functions: It provides its cells with a stable fluid environment. It makes or gets nutrients and other raw materials, distributes them, and disposes of wastes. It has ways to protect itself against injury and attack. It also has the capacity to reproduce, and often it helps nourish and protect its offspring during their early growth and development.

The big picture is this: Each living plant or animal cell engages in metabolic activities that ensure its own survival. But the structural organization and collective activities of cells in tissues, organs, and organ systems sustain the whole body. They work to keep operating conditions in the internal environment within tolerable limits for individual cells—a state called **homeostasis**.

The structural organization of a plant or animal emerges during growth and development and is correlated with basic functions.

Cells, tissues, and organs require, and collectively help maintain, a favorable internal environment. The internal environment consists of all body fluids not inside cells.

The basic functions include maintaining a stable internal environment, securing substances and distributing them through the body, disposing of wastes, protecting the body, reproducing, and often nurturing offspring.

THE NATURE OF ADAPTATION

Defining Adaptation

"Adaptation" is one of those words that have different meanings in different contexts. An individual plant or animal often quickly adjusts its form, function, and behavior. Junipers that germinated in inhospitably windy places are stunted compared to more sheltered individuals of the same species. A sudden clap of thunder may make you lurch the first time you hear it, but you may get used to the sound over time and eventually ignore it. These are *short-term* adaptations to the environment, because they last only as long as the individual does.

Over the long term, an **adaptation** is some heritable aspect of form, function, behavior, or development that improves the odds for surviving and reproducing in a given environment. It is an *outcome* of microevolution —natural selection especially—an enhancement of the fit between the individual and prevailing conditions.

Salt-Tolerant Tomatoes

As a simple example of a long-term adaptation, we can compare the way different tomato species respond to saltwater. Tomatoes evolved in Ecuador, Peru, and the Galápagos Islands. The commercial tomato in grocery stores, *Lycopersicon esculentum*, has eight close relatives in the wild. Mix together ten grams of table salt and sixty milliliters of water, then pour it into a soil-filled container in which an *L. esculentum* plant is growing. Within thirty minutes, the plant will wilt severely (Figure 28.5*a*). Even if soil has only 2,500 parts per million of salt, this species will grow poorly.

Yet the Galápagos tomato (*L. cheesmanii*) survives and reproduces in seawater-washed soils. Researchers

Figure 28.5 (**a**) Severe and rapid wilting of a commercial tomato plant (*Lycopersicon esculentum*) after a drink of salty water. (**b**) Galápagos tomato plant, *L. cheesmanii*.

showed that its salt tolerance is a long-term, heritable adaptation. Gene transfers from the wild species into the commercial species yielded a small, edible tomato. It tolerates irrigation with water that is two parts fresh and one part salty. The hybrid is garnering interest in places where fresh water is scarce and where salts have accumulated in fields used for agriculture.

It may take modification of only a few traits to get salt-tolerant plants. For instance, researchers discovered that revving up a single gene for a sodium/hydrogen ion transporter lets tomato plants use salty water and still bear edible fruits. The fruits are slightly more salty, but the leaves store most of the excess salts.

No Polar Bears in the Desert

You can safely assume that a polar bear (*Ursus maritimis*) is well adapted to the Arctic environment, and that its form and function would be a total flop in a desert (Figure 28.6). You might be able to make some educated guesses about

Figure 28.6 Which adaptations of a polar bear (*Ursus maritimus*) won't help in a desert? Which ones help an oryx (*Oryx beisa*)? For each animal, make a preliminary list of possible structural and functional adaptations relative to the environment. Later, after you finish Unit VI, see how you can expand the list.

why that is so. However, detailed knowledge of its anatomy and physiology might make you view it—or any other animal or plant—with respect. How does a polar bear maintain its internal temperature when it sleeps on ice? How can its muscles function in frigid water? How often must it eat? How does it find food? Conversely, how can an oryx walk about all day in the blistering heat of an African desert? How does it get enough water when there is no water to drink? You will find some answers, or at least ideas about how to look for them, in the next two units of this book.

Adaptation to What?

From the examples just given, you may be thinking it is fairly easy to identify a direct relationship between an adaptation and some aspect of the environment. But the environment in which a trait originally evolved may be very different from the one prevailing now.

Consider the llama, a native of the cloud-piercing peaks of the Andes range paralleling the western coast of South America (Figure 28.7). It can live comfortably 4,800+ meters (16,000 feet) above sea level. Compared to humans living at lower elevations, its lungs have far more air sacs and blood vessels, which formed as it was growing up. A llama heart has larger chambers, so it pumps larger volumes of blood. It does not have to stimulate production of extra blood cells, as people do when they move permanently from low to higher elevations. (The extra cells make the blood "stickier," so the heart has to pump harder.) However, the most publicized adaptation is this: Llama hemoglobin is better than ours at latching onto oxygen. It can pick up oxygen inside the lungs much more efficiently.

Superficially, at least, the oxygen-binding affinity of llama hemoglobin appears to be an adaptation to thin air at high altitudes. But is it? Apparently not.

Llamas belong to the same family as dromedary camels. They share camelid ancestors that evolved in the Eocene grasslands and deserts of North America. Later on, a genetic divergence occurred. The ancestors of camels entered Asia's low-elevation grasslands and deserts by way of a land bridge that later submerged owing to a long-term rise in sea level. The ancestors of llamas moved in a different direction—down the Isthmus of Panama, and on into South America.

Intriguingly, the dromedary camel's hemoglobin also has a high oxygen-binding capacity. So if the trait arose in a shared ancestor, then in what respect has it been adaptive at *low* elevations when it is adaptive at *high* elevations? In this case we can rule out convergent evolution. Why? These animals are very close kin, and their most recent ancestors faced different challenges in environments with different oxygen concentrations.

Figure 28.7 Adaptation to what? A trait is an adaptation to a specific environment. Hemoglobin of llamas, which live at high altitudes, has a high oxygen-binding affinity. But so does hemoglobin of camels, which live at lower elevations.

Who knows why the trait was originally favored? Eocene climates were alternately warm and cool, and hemoglobin's oxygen-binding capacity does go down as temperatures go up. Did it prove adaptive during a long-term shift in the air temperature? Or did the trait have neutral effects at first? For example, what if the mutant gene for the trait became fixed in some ancestral population simply by chance?

Or what if the nonmutated allele interacted closely with another gene that could not be dispensed with? For example, mechanisms that control certain stages of animal development can't be significantly modified without causing major, and usually lethal, disruptions to the basic body plan. As you will read in Section 43.5, molecular details of most genes that code for these mechanisms have been conserved through time.

Use all of these "what-ifs" as a reminder to think critically about the connections between an organism's form and function. Identifying those connections takes a lot of guesswork, research, and experimental tests.

A long-term adaptation is a heritable aspect of form, function, behavior, or development that contributes to the fit between the individual and its environment.

An adaptive trait improves the odds of surviving and reproducing, or at least it did so under conditions that prevailed when genes encoding the trait first evolved.

Observable traits are not always easy to correlate with specific conditions in the organism's environment.

MECHANISMS OF HOMEOSTASIS IN ANIMALS

In preparation for your trek through the next two units, take time now to get a firm grasp of what homeostasis means to survival. Animal physiologists were the first to identify this state and the mechanisms that maintain it, so let's start with a couple of examples of animals.

Like other adult humans, your body contains many trillions of living cells. Each cell must draw nutrients from and dump wastes into the same fifteen liters of fluid, which is less than sixteen quarts. Again, fluid not inside the cells of your body is the extracellular fluid. Much of this is **interstitial fluid**; it fills spaces between cells and tissues. The rest is **plasma**, the fluid portion of blood. Interstitial fluid exchanges many substances with the cells it bathes and also with blood.

Homeostasis, again, is a state in which the body's internal environment is being maintained within some range that its living cells can tolerate. In nearly all animals, three kinds of components interact to maintain it. They are sensory receptors, integrators, and effectors.

Sensory receptors are cells or cell parts that detect forms of energy, such as pressure. Any specific form of energy that a receptor has detected is a **stimulus**. When one chimp kisses another, pressure on its lips changes. Receptors in the lip tissues translate the stimulus into signals that flow to the brain (Figure 28.8).

The brain is an example of an **integrator**, a central command post that pulls together information about stimuli and issues signals to muscles, glands, or both. Muscles and glands—the body's **effectors**—carry out suitable responses. One response to a kiss is flushing with pleasure and kissing back. Of course, a kiss can't continue indefinitely. Doing so would prevent eating, breathing deeply, and various other activities needed to maintain operating conditions in the body.

How does the brain reverse physiological changes induced by the kiss? Receptors can only provide it with information about how things *are* operating. The brain also evaluates information about how things *should be* operating relative to "set points." One example of a set point is a certain concentration of carbon dioxide in the blood coursing through a certain artery. When carbon dioxide's concentration deviates sharply from the set point, the brain initiates actions that will return it to an effective operating range. It does so by way of signals that cause specific effectors in different body regions to increase or decrease certain activities.

Negative Feedback

Feedback mechanisms are important controls that help keep physical and chemical aspects of the body within tolerable ranges. A major type is the **negative feedback mechanism**: An activity is initiated and changes some condition, and when the condition is altered enough, it triggers a response that reverses the change. Figure 28.9 shows one model for such control.

Think of a furnace with a thermostat. A thermostat senses the surrounding air's temperature relative to a preset point on a thermometer built into the furnace's control system. When the temperature falls below the preset point, it sends signals to a switching mechanism that turns on the furnace. When the air becomes heated enough to match the prescribed level, the thermostat signals the mechanism, which shuts off the furnace.

Similarly, feedback mechanisms help keep the core body temperature of chimpanzees, humans, huskies, and many other animals near 37°C (98.6°F) even in hot or cold weather. Thinks of a husky running around on a hot summer day. Its body heats up. Receptors trigger events that slow down the whole dog *and* its cells. The husky searches for shade and flops down under a tree. Moisture from its respiratory system evaporates from the tongue and carries away some body heat with it, as

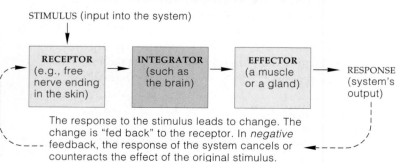

STIMULUS (input into the system)

| RECEPTOR (e.g., free nerve ending in the skin) | → | INTEGRATOR (such as the brain) | → | EFFECTOR (a muscle or a gland) | → | RESPONSE (system's output) |

The response to the stimulus leads to change. The change is "fed back" to the receptor. In *negative* feedback, the response of the system cancels or counteracts the effect of the original stimulus.

Figure 28.8 A kiss. It cannot continue indefinitely without disrupting oxygen flow, food intake, and other physiological events that maintain the internal environment. Negative feedback sets in at the organ level. It reverses the changes brought about by the pressure on the lips.

STIMULUS

The husky is overactive on a hot, dry day and its body surface temperature rises.

| RECEPTORS in skin and elsewhere detect the temperature change. | An INTEGRATOR (the hypothalamus, a brain region) compares input from the receptors against a set point. | Some EFFECTORS (pituitary gland and thyroid gland) trigger widespread adjustments. |

RESPONSE

Temperature of circulating blood starts decreasing.

Many EFFECTORS *carry out specific responses:*

SKELETAL MUSCLES	SMOOTH MUSCLE IN BLOOD VESSELS	SALIVARY GLANDS	ADRENAL GLANDS
Husky rests, starts to pant (behavioral changes).	*Blood carrying metabolically generated heat shunted to skin, some heat lost to surroundings.*	*Secretions from glands increase; evaporation from tongue. Both have a cooling effect, especially on the brain.*	*Output drops, husky is less stimulated.*

Activity of the body in general slows down (behavioral change).

The overall slowdown in activities results in less metabolically generated heat.

Figure 28.9 Homeostatic controls over the internal temperature of a husky's body. The *blue* arrows indicate the main control pathways. The dashed line shows how a feedback loop is completed.

the Figure 28.9 indicates. These control mechanisms and others counter the overheating by curbing the activities that naturally generate metabolic heat and by giving up the body's excess heat to the surrounding air.

Positive Feedback

In some cases, **positive feedback mechanisms** operate. These controls initiate a chain of events that *intensify* change from an original condition, and after a limited time, the intensification reverses the change. Positive feedback is associated with instability in a system. For example, during sexual intercourse, chemical signals from the nervous system of a human female might induce her to make intense physiological responses to stimulation from her partner. Responses may stimulate changes in her partner that stimulate the female more, and so on until she reaches an explosive, climax level of excitation. Now the affected parts of the body return to normal, and homeostasis prevails.

As another example, at the time of childbirth, the fetus exerts pressure on the wall of its mother's uterus. The pressure stimulates the production and secretion of oxytocin, a hormone. Oxytocin causes muscle cells in the wall to contract. Contractions exert pressure on the fetus, which exerts more pressure on the wall, and so on until the fetus is expelled from her body.

What we have been describing is a general pattern of detecting, evaluating, and responding to a continual flow of information about the animal's internal and external environments. During all of the activity, organ systems operate together in astoundingly coordinated fashion. In time you should find yourself asking these questions about their operation:

1. Which physical or chemical aspects of the internal environment are organ systems working to maintain as conditions change?

2. By what means are organ systems kept informed of the various changes?

3. By what means do they process the information?

4. What mechanisms are set in motion in response?

As you will read in Unit VI, the organ systems of most animals are under neural and endocrine control.

Homeostatic control mechanisms help maintain physical and chemical aspects of the body's internal environment within ranges that are most favorable for cell activities.

DOES THE CONCEPT OF HOMEOSTASIS APPLY TO PLANTS?

Plants differ from animals in some important respects. In young plants, for instance, new tissues arise only at the very tips of actively growing roots and shoots. In animal embryos, tissues and organs form all through the body. Also, plants do not respond to stimuli with a centralized integrator, such as a brain. This means that direct comparisons between plants and animals are not always possible.

But plants do have decentralized mechanisms that work to maintain the internal environment and ensure survival for the body as a whole. It is a bit of a leap, but the concept of homeostasis can be applied to them, too. The next two examples will make the point.

Walling Off Threats

Unlike people, trees consist mostly of dead and dying cells. Also unlike people, trees cannot run away from attacks. And when a pathogen infiltrates their tissues, trees cannot unleash an immune system in response, because trees have none. Some trees live in habitats too harsh or too remote for most pathogens, so they have been able to grow, albeit slowly, for thousands of years. Remember the bristlecone pines?

Most trees growing elsewhere can wall off threats to their internal environment by building a fortress of thickened cell walls around wounds. At the same time, they deploy phenols and other compounds that are toxic to invaders. For example, some cells of conifers and other trees secrete resin. A heavy resin flow can saturate and protect bark and wood near the attack site. It also may seep into the soil and litter above the roots. Taken together, plant responses to an attack are called **compartmentalization**.

So potent are some toxins that they also kill cells of the tree itself. As compartments form around injured or infected or poisoned sites, the tree lays down new tissues over them. This works well when the tree acts strongly or is not under massive attack. Drill holes in a tree species that makes a strong response and it walls off the wounds fast. Drill holes in a tree species that makes a moderate response, and it decays lengthwise and somewhat laterally. Drill holes in a species that is a weak compartmentalizer, though, and it will massively decay (Figure 28.10). Even strong compartmentalizers live only so long. If they form too many walls, they shut off the vital flow of water and solutes to living cells.

Bob Tiplady, a plant health specialist, has reported on compartmentalization in response to attack by fungi of the genus *Armillaria*. These fungi destroy shrubs as well as trees in forests and under cultivation throughout the world. They are saprobic decomposers that help to cycle nutrients back to plants. However, they become opportunistic pathogens among trees already stressed by wounds, aging, drought, insects, and air pollution. They infiltrate roots and draw nutrients from narrow zones of actively growing tissues beneath the bark.

In temperate climates, aboveground symptoms of a fungal infection include a gradual decline in growth. Leaves are undersized or yellow, and they often drop prematurely. In summer, the youngest branches may die back or brown quickly and die.

When compartmentalization is strong, an infection can be stopped. Even then, localized decay continues, as in Figure 28.10. An infected tree may die abruptly or slowly, one limb at a time. For instance, if soil holds ample water, the tree might stay green for up to two years after an *Armarillia* infection has decayed most of its roots. Dry spells add to the stress and invite a quick death. That is one reason fungus-infected trees seem to give up abruptly during a prolonged drought.

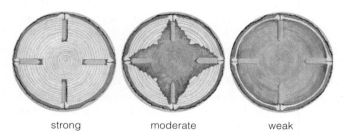

strong   moderate   weak

a Compartmentalization responses

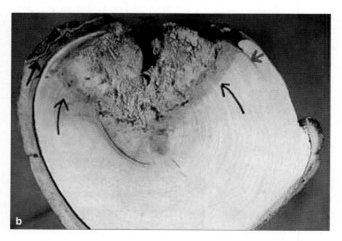

Figure 28.10 (**a**) Pattern of drilling into the stems of three tree species for an experiment to test the effectiveness of compartmentalization. From left to right, the decay patterns (*green*) for three species that are strong, moderate, and weak compartmentalizers. (**b**) Compartmentalized wood rot. Toxic secretions by this tree (*Populus tremuloides*) weakly countered an *Armillaria* attack. A barrier became established (*red* arrow), but the fungus breached it and spread into the wood.

1 A.M.

6 A.M.

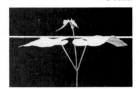
NOON

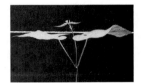
3 P.M.

10 P.M.

c   MIDNIGHT

Sand, Wind, and the Yellow Bush Lupine

Anyone who has tiptoed barefoot across sand near the coast on a hot, dry day has a tangible clue to why few plants grow in it. One of the exceptions is the yellow bush lupine, *Lupinus arboreus* (Figure 28.11*a*).

L. arboreus is a colonizer of soil exposed by fires or abandoned after being cleared for agriculture. It also becomes established along windswept, sandy shores of the Pacific Ocean. Like its relatives in the legume family, it houses nitrogen-fixing symbionts in its roots, which gives it a competitive edge in nitrogen-deficient soils (Chapters 24 and 30).

A big environmental challenge near the beach is the scarcity of fresh water. Leaves of a yellow bush lupine are structurally adapted for water conservation. Each has a surprisingly thin cuticle. But a dense array of fine epidermal hairs projects above the cuticle, especially on the lower leaf surface. Like the hairs on edelweiss, they trap moisture escaping from stomata, and they might also reflect heat. The trapped, moist air slows evaporation and helps keep water inside the leaf.

These leaves make homeostatic responses to the environment. They fold along their length, like the two parts of a clam shell, and resist the moisture-sucking force of the wind. Each folded, hairy leaf is better at holding moisture escaping from stomata (Figure 28.11*b*).

Leaf folding by *L. arboreus* is a controlled response to changing conditions. When winds are strong and the potential for water loss is high, its leaves fold tightly. We find the least-folded leaves near the plant's center or on the side most sheltered from wind. Folding also is a response to heat as well as to wind. When the air temperature is highest during the day, leaves fold at an angle that helps reflect the sun's rays away from their surface. This response minimizes heat absorption.

About Rhythmic Leaf Folding

Just in case you think leaf folding couldn't possibly be a coordinated response, look at Figure 28.11*c*. Like some other plants, this one holds its leaves horizontally in

Figure 28.11 (**a**,**b**) The yellow bush lupine, *Lupinus arboreus*, an exotic species in this habitat by a sandy shore. Introduced into northern California in the early 1900s, it is taking over sand dunes; it outcompetes the native species, to the enormous consternation of conservationists.

(**c**) Observational test of the rhythmic movements of the leaves of a young bean plant (*Phaseolus*). The investigator, physiologist Frank Salisbury, kept this plant in full darkness for twenty-four hours. Its leaves continued to fold and unfold independently of sunrise (6 A.M.) and sunset (6 P.M.).

the day and folds them closer to its stem at night. Keep this plant in full sun or darkness for a few days and it will continue to move its leaves into and out of the "sleep" position, independently of sunrise and sunset. This folding response might help reduce heat loss at night, when air becomes cooler, and so maintain the plant's internal temperature within a tolerable range.

Rhythmic leaf movements are just one example of **circadian rhythm**, a biological activity that is repeated in cycles, each lasting for close to twenty-four hours. Circadian means "about a day." As you will see later, in Chapter 32, a pigment molecule called phytochrome may be part of homeostatic controls over leaf folding.

Homeostatic mechanisms are at work in plants, although they are not governed from central command posts as they are in most animals. Compartmentalization and rhythmic leaf movements are two examples of responses to specific environmental challenges.

COMMUNICATION AMONG CELLS, TISSUES, AND ORGANS

Signal Reception, Transduction, and Response

Reflect on an earlier overview of how adjoining cells communicate, as through plasmodesmata in plants and gap junctions in animals (Section 4.11). Also think back on how free-living *Dictyostelium* cells issue signals to get together and then differentiate into a spore-bearing structure. These amoeboid cells do so in response to dwindling supplies of food—an environmental cue for change (Section 22.12). In large multicelled organisms, cells in one tissue or organ also signal cells in different tissues or organs that are often quite a distance away. They do so in response to changes in the internal and external environments. Their local and long-distance signals call for local and regional changes in metabolic activities, gene expression, growth, and development.

The molecular mechanisms by which cells "talk" to one another evolved early in the history of life, among prokaryotic species. Many persist in the most complex eukaryotic organisms. In many cases, they involve three events—*activation of a receptor, as by reversible binding of a signaling molecule, then transduction of the signal into a molecular form that can operate inside the cell, and then the functional response.*

Most receptors are membrane proteins of the sort shown in Section 5.2. When activated, many change shape, which starts signal transduction. One enzyme often activates many molecules of a different enzyme, which activates many molecules of another kind, and so on in cascading reactions that amplify the signal. In some pathways, small, easily diffusible molecules that are already inside the cytoplasm help broadcast the message through the cell; they are second messengers.

In the next two units, you will come across diverse cases of signal reception, transduction, and response. For now, two simple examples will give you a sense of the kinds of events they set in motion.

Communication in the Plant Body

Have you ever noticed green nubbins along the branch of a young tree? Each holds a community of cells, a meristem, much like tissues of stem cells in your body. They can divide again and again, and the descendants give rise to new tissues and organs. In plants, growth and development of new parts are under the control of genes, signaling molecules, and environmental cues. Incoming light, gravity, seasonal temperature changes, and how long it stays dark at night are such cues.

You might be surprised that growth, development, and survival of a plant require extensive coordination among cells, just as they do in animals. Plant cells, too, communicate with others that are often relatively far away in the plant body. **Hormones** are the main signals for cell plant communication. They initiate changes in cell activities when they dock at suitable receptors.

But hormones do not work in a vacuum. Expression of specific genes in some cells can change a hormone's level in plant tissues or mediate cell sensitivity to it. Expression of many genes is activated or inhibited in response to environmental cues. Chapter 32 describes the main control mechanisms in plants. Here, consider a few genes that control how to make floral organs. These *organ-identity* genes are analogous to homeotic genes, which guide the patterned formation of organs in animals (Sections 43.5 and 43.6).

All flowers are variations on the same pattern of growth and development. Focus on the common wall cress, *Arabidopsis thaliana* (Figure 28.12). It is a valued organism for genetics experiments. This plant is small and has a short generation time; forty-two days after seeds germinate, the new plants have made seeds.

Genetic analysis of *A. thaliana* mutants support an **ABC model** for flowering—that three groups of genes designated *A, B,* and *C* are master switches for floral

petal

carpel

stamen

sepal

Figure 28.12 Observable evidence that meristematic cells at a shoot tip of *Arabidopsis thaliana* were reprogrammed to make a flower instead of more leaves. In (**a**), notice the individual cells making up tissues of the rudimentary structures.

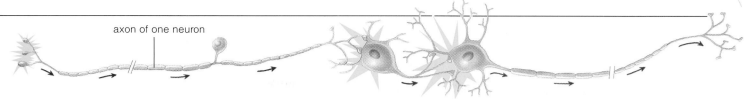

a Reception of environmental signal (pressure) by a sensory neuron in a toe.

b Signal transduced; gates on ion channels across neural membrane open. Rapid ion flow changes the electric and concentration gradients across membrane. Flow reversals self-propagate along length of membrane.

c Signal transduced again; ion flow triggers release of signaling molecules from neuron. This is the response to the signal. Molecules diffuse to another neuron and bind with receptors on it. Binding triggers same transductions in this neuron, and the next, and so on through communication lines from receptor cells to the brain or spinal cord, and on to effectors.

Figure 28.13 (**a–c**) Example of signal reception, transduction, and response in the vertebrate nervous system. A myelin fatty wrapping around the long extensions of these three cells functions in signal propagation. (**d**) Artist's interpretation of the damage caused by an autoimmune attack on myelin sheaths around motor axons. Such attacks cause many disorders, including multiple sclerosis.

development. These groups encode factors governing gene transcription for the products that make sepals, petals, and other structures. They reprogram the mass of rapidly dividing meristematic cells that, until now, were the source of leaves. The descendant cell lineages become arranged in whorls around the shoot's tip, which is a genetically guided floral pattern.

Sepals form as cells in the outer whorls of the floral meristem divide and differentiate. These are the only cells that express *A* genes. Petals form from cells in the whorl inside them, the only cells to express both *A* and *B* genes. *B* and *C* gene interactions switch on activities that result in pollen-producing reproductive structures, the stamens. Cells of the innermost whorl switch on *C* genes only. These cells give rise to a single carpel, the reproductive structure where eggs, then seeds, form.

Another master gene, designated *Leafy*, flips on the ABC switches. And what flips on *Leafy*? At this writing, most evidence is pointing to a steroid hormone.

Communication in the Animal Body

We see elegant cases of signal reception, transduction, and response in cells of the vertebrate nervous system. Chapters 34 and 35 will describe how communication lines of cells called neurons extend through the body. Parts of each neuron are signal input zones; each has a patch of receptor-rich membrane. Other parts, called axons, are typically long extensions with output zones. Chemical signals are released from output zones and diffuse to receptors on a nearby cell (Figure 28.13).

How is a signal transduced? The receptors are on gated ion channels across the plasma membrane. When they bind a signaling molecule, the gates open, and ions flow across the membrane. The distribution of charge across the membrane changes abruptly, which causes ions to flow back across the membrane at an adjacent

patch, and so on in a self-propagating way. When the disturbance reaches the output zone, it triggers release of signaling molecules that act on the next cell.

Many axons have a sheath of **myelin**, a lipid-rich membrane. (Cells called oligodendrocytes make these sheaths by wrapping around the axons like jellyrolls.) A myelin sheath insulates an axon in a way that makes messages travel a lot faster to an output zone. Once in place, it prevents a neuron from sprouting more axons, so it also affects how the developing brain gets wired. It takes twenty years or more before the human brain becomes fully myelinated.

Sometimes the myelin sheaths get damaged. One outcome is *multiple sclerosis*—a gradual loss of brain and spinal cord function. An autoimmune response is made against a myelin component that may have been modified following a viral infection (Figure 28.13*d*).

Plant and animal cells communicate with one another by secreting signaling molecules into extracellular fluid, and selectively responding to signals from other cells.

Cell communication involves mechanisms for receiving signals, transducing them, and bringing about change in metabolism or gene expression in target cells.

Signal transduction commonly involves shape changes in receptors and other membrane proteins as well as activation of enzymes and second messengers inside the cytoplasm.

RECURRING CHALLENGES TO SURVIVAL

Plants and animals have such spectacularly diverse body plans that we sometimes forget how much they have in common. How often do we even think of the connections between, say, Tina Turner and a tulip (Figure 28.14)? Yet the connections are there, and they start with what individual cells in each multicelled body are doing. Nearly all of the cells are working in the best interest of the whole body—*but the whole body is also working in the best interest of individual cells.*

Constraints on Gas Exchange

Tina's cells, and the tulip's, depend on gas exchanges across their plasma membrane. Like most multicelled heterotrophs, Tina has a fine way to get oxygen to the plasma membrane of each one of her aerobically respiring cells and take away their carbon dioxide wastes. Like most autotrophs, a tulip plant can get carbon dioxide to its photosynthetic cells and remove their oxygen by-products. It also delivers oxygen to aerobically respiring cells throughout the body, as in roots. Both organisms have specific body structures that help individual cells by acquiring, transporting, exchanging, and getting rid of gases.

These structures evolved in response to certain environmental parameters. Remember **diffusion**, the net movement of like molecules or ions of a substance from a region where they are more concentrated to a region where they are less concentrated? Once gases reach a cell's plasma membrane, they diffuse across it. Which way? That depends on the direction of the concentration gradient for each one. And which body parts help keep each gradient pointed in the right direction? The questions will lead you to the stomata of leaves (Chapter 30) and to the respiratory and circulatory systems of animals (Chapters 38 and 40).

Figure 28.14   What do these organisms have in common besides their good looks?

Requirements for Internal Transport

As you know, metabolic reactions are amazingly fast. If it takes too long for substances to diffuse through the body or to and from the surface, the body will shut down. That is one of the reasons cells and multicelled organisms have the sizes and shapes that they do. As they grow, their volume expands in three dimensions (say, length, width, and depth), but the surface increases in two dimensions only. And that is the essence of the **surface-to-volume ratio**. If a body were to become one huge unbroken mass, it would not have enough surface area for fast, efficient exchanges with the environment.

When a body or body part is thin, as it is for the lily pads and flatworm shown at right, substances can easily diffuse between its individual cells and the environment. But the farther individual cells are from an exchange point with the environment, the more they depend on systems of rapid internal transport.

Most plants and animals have vascular tissues, which transport substances to and from cells. In plants, soil water and mineral ions are distributed to aboveground parts through xylem. The photosynthetic products made in leaves are distributed through phloem. Each leaf vein shown in Figure 28.15a contains long strands of xylem and phloem.

In large animals, too, vascular tissues extend from an interface with the environment to individual cells. Each time Tina belts out a song, she must swiftly move oxygen into her lungs and carbon dioxide out of them. Part of her vascular system threads intricately through lung tissue, where gases are exchanged. From there, large-diameter vessels swiftly transport oxygen to all other regions, then branch into tiny capillaries (Figure 28.15b). Blood flow slows here, so diffusion proceeds at a suitable pace between blood, interstitial fluid, and cells. The capillaries branch into transport tubes that deliver carbon dioxide to the lungs, where it is exhaled and where oxygen is again picked up (Chapter 40).

As in plants, that vascular system also transports nutrients and water. Unlike plants, the one in animals interconnects with other organ systems that have roles in maintaining a more complex internal environment. One of these, the immune system, uses the vascular system as highways for moving white blood cells and chemical weapons to tissues that are under siege. Another organ system specializes in regulating the concentration and volume of the internal environment.

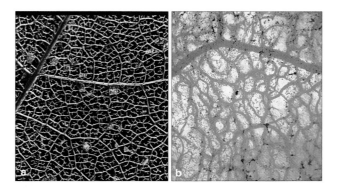

Figure 28.15 Transport lines to and from cells: (**a**) Veins visible in a decaying leaf, and (**b**) human veins and capillaries.

Figure 28.16 Protecting body tissues from predation: (**a**) Cactus spines. (**b**) Quills of a porcupine (*Erethizon dorsatum*).

Maintaining a Solute–Water Balance

Plants and animals continually gain and lose water and solutes. They regularly produce metabolic wastes. Given all the inputs and outputs, how is the volume and composition of their internal environment kept within a tolerable range? Plants and animals differ hugely in this respect. Yet we still can find common responses among them at the molecular level.

Substances tend to follow concentration gradients as they move into and out of the body. They do so as they move to and from one body compartment into another. At such interfaces, we find sheetlike tissues in which cells are engaged in **active transport**, the pumping of specific substances *against* the gradients.

In roots, active transport mechanisms help control which solutes move into the plant. In leaves, they help control water loss and gas exchange by making stomata open only at certain times. In animals, we find such mechanisms in kidneys and many other organs. As you will see again and again in the next two units, *active and passive transport mechanisms help maintain metabolism and the internal environment by continually adjusting the kinds, amounts, and directional movements of substances.*

Requirements for Integration and Control

We could introduce you to many other structural and functional similarities between plants and animals. But let's sign off with the cells of certain tissues, which release the signaling molecules that coordinate and integrate activities of the body as a whole. Sections 28.4 and 28.5 gave a few examples. These are the signals that respond to changes in the internal and external environments. Different kinds guide events by which the body grows, develops, maintains, and often prepares for reproduction—events that will be our focus in chapters to come.

On Variations in Resources and Threats

Beyond the common challenges are resources and dangers that differ among habitats. A **habitat** is the place where individuals of a species normally live.

What are the physical and chemical characteristics of the habitat? Is water plentiful, with suitable kinds and amounts of solutes? Is the habitat rich or poor in nutrients? Is it sunlit, shady, or dark? Is it warm or cool, hot or icy, windy or calm? How much do outside temperatures change from day to night? Are seasonal changes slight or pronounced?

And what about the biotic (living) components of the habitat? What kinds of producers, predators, prey, pathogens, and parasites live there? Is competition for resources and reproductive partners fierce among others of the species? These are the kinds of variables that promote diversity in anatomy and physiology.

Even in all that diversity, we often can find similar responses to similar environmental challenges. Think about the sharp spines of a cactus and sharp quills of a porcupine (Figure 28.16). These specialized structures function effectively as deterrents to most predators. Both are derived from specialized epidermal cells. In both cases, vascular tissues and other body parts helped nurture those cells. And by contributing to the formation of defensive structures at the body surface, individual epidermal cells help protect the whole body against an environmental threat.

Plant and animal cells function in ways that help ensure survival of the body as a whole. At the same time, tissues and organs that make up the body function in ways that help ensure survival of individual living cells.

The connection between each cell and the body as a whole is evident in the requirements for—and contributions to—gas exchange, nutrition, internal transport, stability in the internal environment, and defense.

SUMMARY
Gold indicates text section

1. Anatomy is the study of body form. Physiology is the study of body functioning in the environment. The structure of most body parts correlates with current or past functions. *CI, 28.1*

2. A plant or animal's structural organization emerges during growth and development. Each cell in its body performs metabolic functions that ensure its survival. Cells are organized in tissues, organs, and often organ systems, which function in coordinated ways to ensure survival of the whole organism. *28.1, 28.6*

3. Long-term adaptations are specific heritable traits that evolved in the past and continue to be effective, or at least neutral, under prevailing conditions. The traits are specific aspects of the form, function, behavior, or development of the body. *28.2*

4. Tissues, organs, and organ systems work together to maintain a stable internal environment (extracellular fluid) needed for individual cell survival. At homeostasis, conditions in the internal environment are balanced at levels most favorable for cell activities. *28.1, 28.3, 28.4*

5. Feedback controls help maintain internal operating conditions. Example: By negative feedback, a change in some condition, such as body temperature, triggers a response that results in reversal of the change. *28.3*

6. Cells of multicelled organisms communicate directly with one another. They communicate with cells some distance away, in different tissues and organs. They secrete signaling molecules into extracellular fluid and selectively respond to signals themselves. *28.5*

7. Cell communication involves reception of signals, transduction of them, and a response, such as a change in metabolism, in gene expression, or in development. Signal transduction often involves shape changes in receptors and other membrane proteins. It affects the activity of enzymes and often of second messengers in the cytoplasm of the target cells. *28.5*

Review Questions

1. Define tissue, organ, and organ system. *28.1*

2. Distinguish between growth and development. *28.1*

3. What does the term internal environment mean? *28.1*

4. Define long-term adaptation. Is it a process of evolution or an outcome of microevolutionary processes? *28.2*

5. Define homeostasis and give an example of how certain mechanisms work to maintain it. *28.3, 28.4, 28.6*

6. Briefly define the receptors, integrators, and effectors in the animal body and state how they interact. *28.3*

7. Do plants have decentralized homeostatic mechanisms? *28.4*

8. Describe a form of communication between two or more plant or animal tissues. *28.5*

Self-Quiz
ANSWERS IN APPENDIX III

1. An increase in the number, size, and volume of plant cells or animal cells is called _____ .
   a. growth   c. differentiation
   b. development   d. all of the above

2. The internal environment consists of _____ .
   a. all body fluids   c. all body fluids outside cells
   b. all fluids in cells   d. interstitial fluid

3. As basic functions, a plant or animal must _____ .
   a. maintain a stable internal environment
   b. get and distribute water and solutes through the body
   c. dispose of wastes
   d. defend the body
   e. all of the above

4. Cell communication always involves signal _____ .
   a. reception   c. response
   b. transduction   d. all of the above

5. Match the terms with their most suitable description.
   ____ physiology   a. study of how body parts function
   ____ circadian rhythm b. signaling molecule
   ____ homeostasis   c. 24-hour or so cyclic activity
   ____ hormone   d. stable internal environment
   ____ negative feedback e. activity changes some condition, change causes its own reversal

Critical Thinking

1. Reflect on prevailing conditions in a desert in New Mexico or Arizona; on the floor of a shady, moist forest in Georgia or Oregon; and in Alaska's arctic tundra. Consider the kinds of plants and animals living in each environment. Now "design" a plant or animal that might function even more efficiently in each place. Be sure to consider how its basic requirements, such as the acquisition of water and nutrients, will be met.

2. Refer to Section 28.2. Set up class teams and see which team can come up with the longest list of the most plausible answers to the questions asked in the caption to Figure 28.6.

Selected Key Terms

ABC model *28.5*	integrator *28.3*
active transport *28.6*	internal environment *28.1*
adaptation, long-term *28.2*	interstitial fluid *28.3*
anatomy *CI*	myelin *28.5*
circadian rhythm *28.4*	negative feedback *28.3*
compartmentalization *28.4*	organ *28.1*
diffusion *28.6*	organ system *28.1*
development *28.1*	physiology *CI*
division of labor *28.1*	plasma *28.3*
effector *28.3*	positive feedback *28.3*
growth *28.1*	sensory receptor *28.3*
habitat *28.6*	stimulus *28.3*
homeostasis *28.1*	surface-to-volume ratio *28.6*
hormone *28.5*	tissue *28.1*

Readings

Hochachka, P. and G. Somero. 2002. *Biochemical Adaptation: Mechanism and Process in Physiological Evolution.* New York: Oxford University Press. Integrative look at how cellular systems are adapted to environmental stresses.

On-Line readings at Student Guide for InfoTrac:
www.brookscole.com/biology

V Plant Structure and Function

The sacred lotus, *Nelumbo nucifera*, busily doing what its ancestors did for well over 100 million years—flowering spectacularly during the reproductive phase of its life cycle.

29

PLANT TISSUES

Plants Versus the Volcano

On a clear spring day in 1980, in a richly forested region of the Cascade Range of southwestern Washington, Mount Saint Helens erupted and 500 million metric tons of ash blew skyward. Within minutes, shock waves from the blast had flattened or incinerated hundreds of thousands of tall, mature trees growing near the mountain's northern flanks.

Rivers of hot volcanic ash surged down the slopes at rates exceeding forty-four meters per second. They turned into nightmarish rivers of cementlike mud when intense heat from the blast melted and released more than 75 billion liters of water that had been locked up in the mountain's snowfields and glacial ice.

In one mind-numbing moment, about 40,500 hectares—100,000 acres—of magnificent forests dominated by hemlock and Douglas fir had been transformed into barren sweeps of land (Figure 29.1a,b). In the aftermath of the violent eruption, we gained stunning insight into what the world must have looked like long ago, before the first plants started to colonize habitats on land.

Yet it did not take long for existing plants to move back into habitats that their ancestors had claimed. In less than a year's time, the seeds of a variety of flowering plants, including fireweed and blackberry, sprouted near the grayed trunks of fallen trees around Mount Saint Helens.

Before ten years had passed, young willows and alders took hold near riverbanks, and low shrubs cloaked the land (Figure 29.1c). In time they afforded pockets of shade, which favored germination of seeds of the slower growing but ultimately dominant species, the hemlocks and Douglas fir (Figure 29.1d). In less than a century, the forest will be as it once was.

Figure 29.1 (**a**,**b**) Grim reminder of what the world would be like without plants: aftermath of the violent eruption of Mount Saint Helens in 1980. Nothing remained of the forest that surrounded this Cascade volcano. (**c**) In less than a decade, seed-bearing vascular plants were making a comeback. (**d**) Twelve years after the eruption, the seedlings of a dominant species, Douglas fir (*Pseudotsuga menziesii*), were starting to reclaim the land.

With this example, we open a unit dedicated to the seed-bearing vascular plants, with emphasis on the flowering types. In terms of distribution and diversity, they are the most successful plants on Earth.

This first chapter is an overview of plant tissues and body plans. Next, Chapter 30 explains how the seed-bearing plants absorb and distribute water and mineral ions, conserve water, and distribute organic substances among roots, stems, and leaves. Chapters 31 and 32 take a look at their patterns of growth, development, and reproduction. As you will see, their structure and physiology (that is, the functioning of the plant body) help them survive sometimes hostile conditions on land—even momentary takeovers by volcanoes.

d

Key Concepts

1. Angiosperms (flowering plants) and, to a lesser extent, gymnosperms are groups that now dominate the plant kingdom. All are seed-bearing vascular plants. They have complex aboveground shoot systems of stems, leaves, and reproductive parts. Most species have complex root systems that grow downward and outward through soil.

2. We find three major tissue systems in seed-bearing vascular plants. A ground tissue system makes up the bulk of the plant body. A vascular tissue system distributes water, dissolved mineral ions, and the products of photosynthesis. A dermal tissue system covers and protects plant surfaces.

3. Simple plant tissues—parenchyma, collenchyma, and sclerenchyma—are each composed of no more than one type of cell.

4. Complex plant tissues incorporate two or more types of cells. Xylem and phloem, which are vascular tissues, are like this. So are the dermal tissues called epidermis and periderm.

5. Plants lengthen and thicken only through mitotic cell divisions and the accompanying cell growth at meristems. At these localized regions of the plant body, rapid divisions of undifferentiated cells give rise to all specialized cell lineages that form mature tissues.

6. Each growing season, shoots and roots lengthen. The lengthening, called primary growth, originates only at apical meristems in shoot and root tips.

7. For many plant species, shoots and roots also thicken during the growing season. Typically, lateral meristems inside shoots and roots give rise to an increase in diameter, which is called secondary growth. Wood is one outcome of secondary growth.

OVERVIEW OF THE PLANT BODY

Earlier, in Chapter 23, we surveyed representatives of the 295,000 known species of plants. Even that sprint through diversity revealed why no one species can be used as a typical example of plant body plans. When we hear the word "plant," however, we usually think of well-known species of seed-bearing vascular plants—gymnosperms (including pine trees) and angiosperms (flower-producing plants, such as roses, corn, cactuses, and elms). With 260,000 species, angiosperms dominate the plant kingdom. Its major groups are **magnoliids, eudicots** (true dicots) and **monocots** (Section 23.8). We focus here on the true dicots and monocots.

Shoots and Roots

Many flowering plants have a body plan similar to that shown in Figure 29.2. Aboveground are the **shoots**: stems, leaves, flowers (reproductive shoots), and other structures. Stems offer structural support for upright growth, and some of its tissues also conduct water and solutes. Upright growth gives photosynthetic cells in young stems and leaves favorable exposure to sunlight. **Roots** are specialized structures that most often grow downward and outward through soil. An underground root system absorbs water and dissolved minerals, and it typically anchors the aboveground parts. A root also stores food, then releases it as required for its own cells and for distribution to living cells aboveground.

Three Plant Tissue Systems

Stems, branches, leaves, and roots all consist of three major tissue systems (Figure 29.2). The **ground tissue system** serves basic functions, such as food and water storage. The **vascular tissue system** has two different tissues that distribute water and solutes. The **dermal tissue system** covers and protects plant surfaces.

Some tissues in each system are simple, in that they have only one type of cell. Parenchyma, collenchyma, and sclerenchyma are in this category. Other tissues are complex, with organized arrays of two or more types of cells. Xylem, phloem, and epidermis are like this.

The next sections describe the tissue organization of shoots and roots. You may find it easier to follow them by studying Figure 29.3. It reviews different ways in which botanists cut tissue specimens from plants.

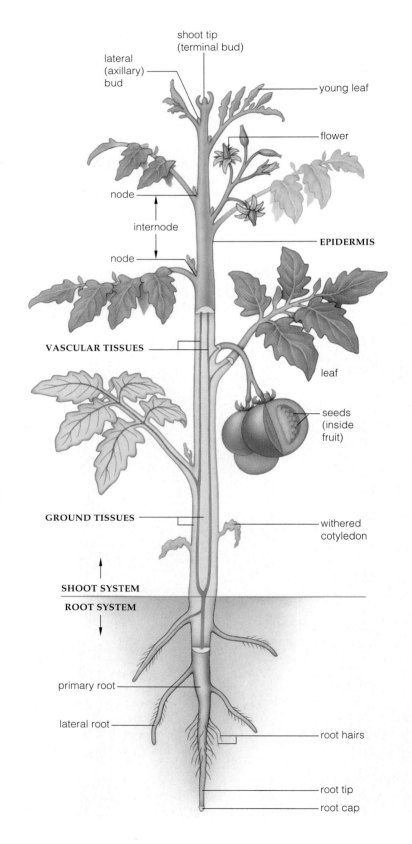

Figure 29.2 Body plan for a tomato plant (*Lycopersicon*). Its vascular tissues (*purple*) conduct water, dissolved minerals, and organic substances. They thread through ground tissues that make up most of this plant. Dermal tissue (epidermis in this case) covers surfaces of the root and shoot systems.

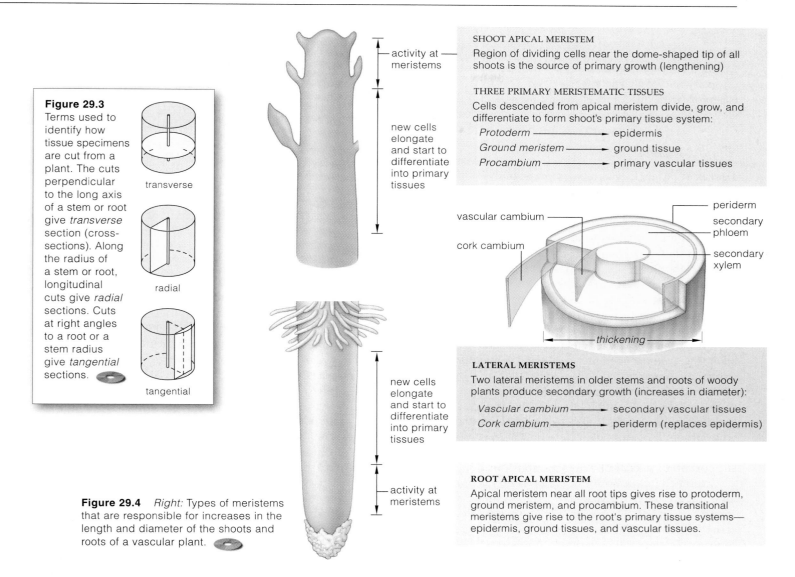

Figure 29.3
Terms used to identify how tissue specimens are cut from a plant. The cuts perpendicular to the long axis of a stem or root give *transverse* section (cross-sections). Along the radius of a stem or root, longitudinal cuts give *radial* sections. Cuts at right angles to a root or a stem radius give *tangential* sections.

transverse

radial

tangential

activity at meristems

new cells elongate and start to differentiate into primary tissues

new cells elongate and start to differentiate into primary tissues

activity at meristems

Figure 29.4 *Right:* Types of meristems that are responsible for increases in the length and diameter of the shoots and roots of a vascular plant.

SHOOT APICAL MERISTEM
Region of dividing cells near the dome-shaped tip of all shoots is the source of primary growth (lengthening)

THREE PRIMARY MERISTEMATIC TISSUES
Cells descended from apical meristem divide, grow, and differentiate to form shoot's primary tissue system:

Protoderm ——————► epidermis
Ground meristem ————► ground tissue
Procambium —————► primary vascular tissues

vascular cambium

cork cambium

periderm

secondary phloem

secondary xylem

thickening

LATERAL MERISTEMS
Two lateral meristems in older stems and roots of woody plants produce secondary growth (increases in diameter):

Vascular cambium ————► secondary vascular tissues
Cork cambium —————► periderm (replaces epidermis)

ROOT APICAL MERISTEM
Apical meristem near all root tips gives rise to protoderm, ground meristem, and procambium. These transitional meristems give rise to the root's primary tissue systems—epidermis, ground tissues, and vascular tissues.

Where Do Plant Tissues Originate?

Different plant tissues become active at different times during the growing season. Most growth proceeds at **meristems**, which are localized regions of dividing cells (Figure 29.4). In other regions, cellular descendants of meristems are maturing or have already matured.

Apical meristems, in the tips of shoots and roots, are where plant parts start to lengthen. Populations formed here develop into protoderm, ground meristem, and procambium. These are immature forms of the primary tissues—epidermis, ground tissue, and vascular tissues respectively. Taken as a whole, the *lengthening* of stems and roots represents the plant's primary growth.

Also during a growing season, the older stems and roots of many plants thicken. Increases in girth start with *lateral* meristems, each a cylindrical array of cells that forms in stems and roots. One lateral meristem, the **vascular cambium**, produces secondary vascular

tissues. The other, **cork cambium**, produces a sturdier covering (periderm) that replaces epidermis. Taken as a whole, the *thickening* of stems and roots represents secondary growth.

Vascular plants have stems that support upright growth and conduct substances, leaves that function in photosynthesis, shoots specialized for reproduction, and other structures. They also have roots that absorb water and solutes. Roots often anchor aboveground parts and store food.

A ground tissue system makes up most of the young plant body. A vascular tissue system distributes water, dissolved ions, and photosynthetic products through it. A dermal tissue system covers and protects plant surfaces.

Shoots and roots lengthen (put on primary growth) when their apical and primary meristems are active. In many plants, older stems and roots also thicken (add secondary growth) when lateral meristems called vascular cambium and cork cambium are active.

TYPES OF PLANT TISSUES

We turn now to an overview of the organization and functions of plant tissues. Simple tissues are composed of one type of cell. The vascular and dermal tissues are complex, with different cell types. Figures 29.5 through 29.9 show examples from these tissue categories.

Simple Tissues

Tissues of **parenchyma** make up most of the soft, moist, primary growth of roots, stems, leaves, flowers, and fruits. Most parenchyma cells are pliable, thin-walled, and many-sided. Mature cells are alive and can still divide, often to heal wounds. In leaves, mesophyll is a photosynthetic parenchyma, and air spaces between its cells enhance gas exchange. Parenchyma also has roles in storage, secretion, and other specialized tasks. The vascular tissue systems also contain parenchyma cells.

Collenchyma provides flexible support for primary tissues. Patches or cylinders of its living, usually long cells often support a lengthening stem and form leaf stalk ribs. In their unevenly thickened cell walls, pectin (a pliable polysaccharide) glues cellulose fibrils together.

Most cells in **sclerenchyma** have lignin-impregnated thick walls. Lignin stiffens these walls and gives them compressive strength. It also resists fungal attacks and waterproofs the walls of water-conducting cells. Land plants could not have evolved without its mechanical support and water transport functions (Section 23.1).

Sclerenchyma cells are fibers or sclereids. *Fibers* are long, tapered cells in vascular tissue systems of some stems and leaves (Figure 29.7a). Fibers flex, twist, and resist stretching. We use certain kinds to make cloth, rope, paper, and other valued commodities. *Sclereids* are stubbier cells. Think of a hard seed coat, coconut shell, or peach pit or a pear's gritty texture; sclereids are the source of such features (Figure 29.7b).

Figure 29.5 Locations of simple tissues and complex tissues in one kind of plant stem, transverse section.

epidermis
collenchyma
sclerenchyma
parenchyma

xylem
phloem

Complex Tissues

VASCULAR TISSUES   Two vascular tissues—xylem and phloem—distribute substances through plants. Fibers and parenchyma often sheath their conducting cells.

Xylem conducts water and dissolved mineral ions. It also helps mechanically support plants. Figure 29.8a,b shows examples of its conducting cells. The cells, *vessel members* and *tracheids*, are dead at maturity, and their lignified walls interconnect. Collectively, the cell walls form water-conducting pipelines and strengthen plant parts. Water flows into and out of the adjoining cells through numerous pits in the cell walls.

Phloem conducts sugars and other solutes. Its main conducting cells, called *sieve-tube members*, are alive at maturity (Figure 29.8c). They form tubes that connect at openings in their side walls and end walls. Sugars made by photosynthetic cells in leaves are loaded into sieve-tube members with the help of specialized, living parenchyma cells called *companion cells*. Sugars moving through the phloem pipelines are unloaded in regions where cells are growing or storing food. The chapter to follow describes this process.

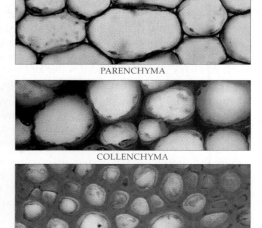

PARENCHYMA

COLLENCHYMA

SCLERENCHYMA

Figure 29.6 Three examples of simple tissue from the stem of a sunflower plant (*Helianthus*), cross-section. Parenchyma makes up the bulk of the plant body. Collenchyma and sclerenchyma help support and also strengthen plant parts.

FIBERS

thick, lignified secondary wall

b

SCLEREIDS

Figure 29.7 Two different kinds of sclerenchyma. (**a**) Strong fibers from flax stems. Compare Figure 4.27a,c. (**b**) From a pear, the sclereids called stone cells, cross-section.

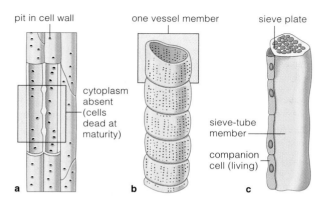

pit in cell wall · one vessel member · sieve plate

cytoplasm absent (cells dead at maturity)

sieve-tube member

companion cell (living)

a · b · c

Figure 29.8 From xylem, portions of (**a**) tracheids and (**b**) a vessel. Pipelines made of such cells conduct water and dissolved ions. (**c**) One type of phloem cell. Long tubes of many cells conduct sugars and other organic compounds.

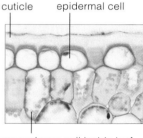

cuticle · epidermal cell

parenchyma cell inside leaf

Figure 29.9 Light micrograph of a section through the upper surface of a kaffir lily leaf. The plant cuticle is made of secretions from epidermal cells. Inside the leaf are many photosynthetic parenchyma cells.

DERMAL TISSUES A dermal tissue system, **epidermis**, covers surfaces of primary plant parts. In most plants, it is mainly a single layer of unspecialized cells. Waxes and cutin, a fatty substance, coat outward-facing cell walls. The surface coating is a **cuticle**, which helps the plant conserve water and in some cases resists attacks by microorganisms (Figure 29.9).

Stem and leaf epidermis contains many specialized cells. For instance, pairs of guard cells change shape in response to changing conditions. As they do, a gap—or **stoma** (plural, stomata)—closes or opens between them. The next chapter looks at how stomata work as control points for the movement of water vapor, oxygen, and carbon dioxide across the epidermis. Periderm replaces epidermis in stems and roots with secondary growth. As you will see, it includes dead cork cells with walls heavily impregnated with suberin, a fatty substance.

Dicots and Monocots—Same Tissues, Different Features

True dicots (eudicots) include most trees and shrubs other than conifers, such as maples, roses, cacti, and beans. Palms, lilies, orchids, ryegrass, wheat, corn, and

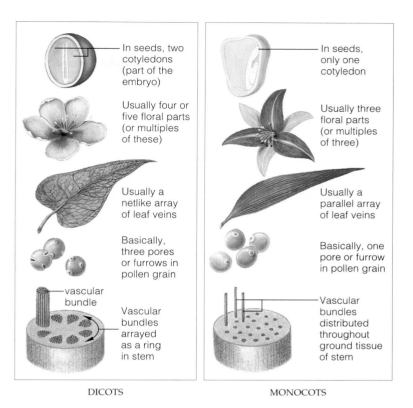

In seeds, two cotyledons (part of the embryo)

Usually four or five floral parts (or multiples of these)

Usually a netlike array of leaf veins

Basically, three pores or furrows in pollen grain

vascular bundle

Vascular bundles arrayed as a ring in stem

In seeds, only one cotyledon

Usually three floral parts (or multiples of three)

Usually a parallel array of leaf veins

Basically, one pore or furrow in pollen grain

Vascular bundles distributed throughout ground tissue of stem

DICOTS · MONOCOTS

Figure 29.10 Comparison of some defining features of the true dicots (eudicots) and monocots. Both classes of flowering plants consist of the same simple and complex tissues, but their body parts show some differences in structural organization.

sugarcane are familiar monocots. Dicots and monocots are similar in structure and function but differ in some distinctive ways. Dicot seeds have two cotyledons and monocot seeds have only one. Cotyledons are leaflike structures commonly known as seed leaves. They form in seeds as part of a plant embryo, and they store or absorb food for it. After a seed germinates, cotyledons wither, and new leaves grow and start to make food. Figure 29.10 shows other differences between dicots and monocots.

Most of the plant body (ground tissue system) consists of parenchyma, collenchyma, and sclerenchyma. Each of these simple tissues is composed of only one type of cell.

Xylem and phloem are vascular tissues. In xylem, pipelines made of tracheids and vessel members conduct water and dissolved ions. In phloem, sieve-tube members interact with companion cells to distribute organic compounds.

Of two dermal tissues, epidermis covers the surfaces of the primary plant body. Periderm replaces epidermis on plant parts with extensive secondary growth.

Dicots and monocots consist of the same tissues, but each has some of the tissues organized in distinctive ways.

PRIMARY STRUCTURE OF SHOOTS

How Do Stems and Leaves Form?

Next time you or a friend eats a bundle of bean sprouts or alfalfa sprouts, pull one aside to look at its structure. That seedling started forming while it was still inside a seed coat. It already has a primary root and shoot. In the primary shoot's tip, apical meristem and its descendant tissues are laying out orderly frameworks for the stem primary structure (Figure 29.11). Below the shoot tip, tissues become specialized as cells divide at different rates in prescribed directions, and differentiate in size, shape, and function. These tissues make up distinctive stem regions, leaves, and lateral (axillary) buds from which lateral shoots develop. Lateral shoots give rise to side branches and reproductive structures.

Briefly, as a typical shoot lengthens, bulges of tissue develop along the sides of apical meristem. Each bulge is one immature leaf (Figures 29.11 and 29.12). While growth continues, the stem lengthens between tier after tier of new leaves. Each part of the stem where one or more leaves are attached is a node. As shown in Figure 29.2, the stem region between two successive nodes is an internode. Buds develop in the upper angle where leaves attach to the stem. Each **bud** is an undeveloped shoot of mostly meristematic tissue, often protected by bud scales (modified leaves). Buds give rise to stems, leaves, and flowers in ways described in Section 32.5.

Internal Structure of Stems

While the primary plant body of a monocot or dicot is forming, the ground, vascular, and dermal tissues of its stems become organized in distinctive ways. Most often, primary xylem and phloem develop inside the same sheath of cells, as **vascular bundles**. The bundles are multistranded cords threading lengthwise through the ground tissue system of primary and lateral shoots.

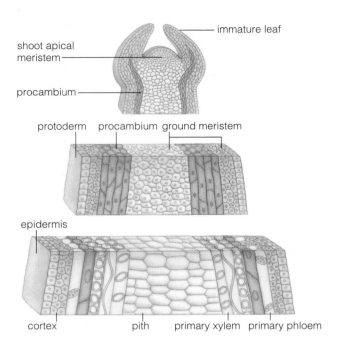

Figure 29.11 Successive stages in primary growth, starting with activity at shoot apical meristem of a typical dicot and continuing at primary meristem tissues derived from it. Notice the progressive differentiation of most tissue regions.

They commonly develop in two genetically dictated patterns. In most dicot stems, long bundles form a ring that divides the ground tissue into a cortex and pith (Figure 29.13a). A stem's **cortex** is the region between the vascular bundles and the epidermis. Its **pith** is the center of the stem, inside the ring of vascular bundles. Ground tissue of the plant's roots becomes similarly divided, into root cortex and pith. A different pattern is common inside the stems of most monocots and some

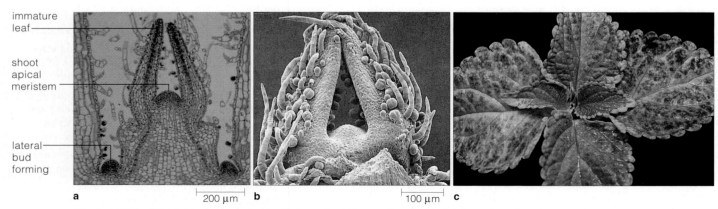

Figure 29.12 (**a**) Light micrograph of a *Coleus* shoot tip, cut longitudinally through its center. (**b**) Scanning electron micrograph of its surface. (**c**) New *Coleus* leaves.

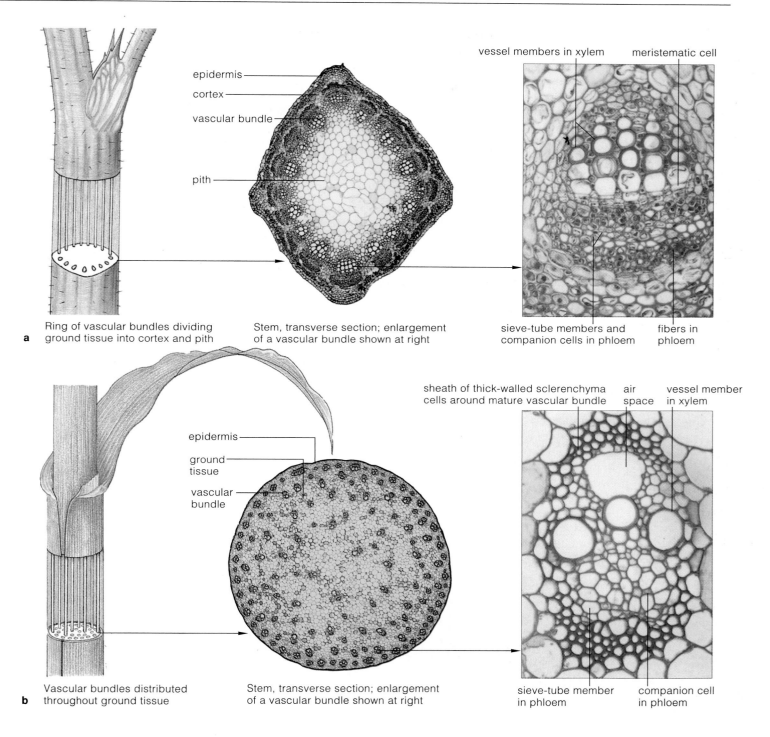

a Ring of vascular bundles dividing ground tissue into cortex and pith

epidermis
cortex
vascular bundle
pith

Stem, transverse section; enlargement of a vascular bundle shown at right

vessel members in xylem meristematic cell

sieve-tube members and companion cells in phloem fibers in phloem

b Vascular bundles distributed throughout ground tissue

epidermis
ground tissue
vascular bundle

Stem, transverse section; enlargement of a vascular bundle shown at right

sheath of thick-walled sclerenchyma cells around mature vascular bundle air space vessel member in xylem

sieve-tube member in phloem companion cell in phloem

Figure 29.13 Internal organization of cells and tissues inside the stems from a dicot and a monocot. (**a**) Part of a stem from alfalfa (*Medicago*), a dicot. In many species of dicots and conifers, the vascular bundles develop in a more or less ringlike array in the ground tissue system, as shown here. The portion of the ground tissue between the ring and the surface of the stem is the cortex. The portion enclosed within the ring is the pith. (**b**) Part of a stem from corn (*Zea mays*), a monocot. In most monocots and some nonwoody dicots, vascular bundles are scattered through the ground tissue, as shown.

dicots. The long vascular bundles inside the stem are scattered throughout its ground tissue (Figure 29.13*b*). How substances are conducted through such vascular systems is a topic of the next chapter.

Shoot apical meristems give rise to the primary plant body, which develops a distinctive internal structure (as in the pattern in which its vascular bundles are arranged).

A CLOSER LOOK AT LEAVES

Similarities and Differences Among Leaves

Every **leaf** that forms is a metabolic factory equipped with many photosynthetic cells. Yet leaves vary greatly in size, shape, surface details, and internal structure. A duckweed leaf is no more than 1 millimeter (0.04 inch) across; leaves of one palm (*Attalea*) are 12 meters (40 feet) across. Different leaves are shaped like needles, blades, spikes, cups, tubes, and feathers. They differ hugely in color, odor, and edibility; many form toxins. Leaves of birches and other species of *deciduous* plants wither and drop away from stems with the approach of winter. Leaves of camellias and other *evergreen* plants also drop, but not all at the same time.

A typical leaf has a flat blade, as in Figures 29.14*a* and 29.15. It has a stalk (petiole) that attaches it to the stem. *Simple* leaves are undivided, but many are lobed. *Compound* leaves have blades divided into leaflets, all oriented in the same plane. Leaves of most monocots, such as ryegrass and corn, are flat surfaced like a knife blade. The blade's base encircles and sheaths the stem.

Leaves of most species are thin, with a high surface-to-volume ratio. Their flat surface grows and orients itself perpendicular to light. A leaf often projects from a stem in patterns that minimize shading of other leaves. For example, a clover leaf stalk is attached to the stem at right angles to its neighbors (Figure 29.15*c*).

Such leaf adaptations intercept as much of a plant's energy source —sunlight—as possible. They also help oxygen diffuse out and carbon dioxide diffuse in. And when a leaf

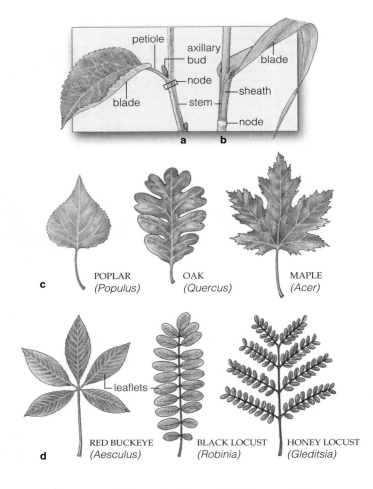

Figure 29.14 Common leaf forms of (**a**) dicots and (**b**) monocots. Examples of (**c**) simple leaves and (**d**) compound leaves.

is thick, you can assume it belongs to a succulent or some other plant of arid habitats and serves in water storage as well as photosynthesis. Another example: The leaves of many desert plants orient themselves parallel with the sun's rays to reduce heat absorption to tolerable levels.

Leaf Fine Structure

In its fine structure, too, each leaf is adapted to intercept energy from the sun's rays and promote gas exchange. In addition, many have distinctive surface specializations.

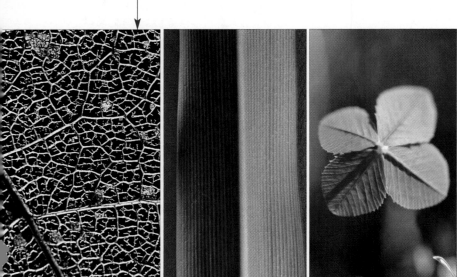

Figure 29.15 (**a**) Decaying dicot leaf, with its netlike veins. (**b**) Parallel veins of a monocot leaf (*Agapanthus*). (**c**) Leaf orientation of a four-leaf clover (*Trifolium*).

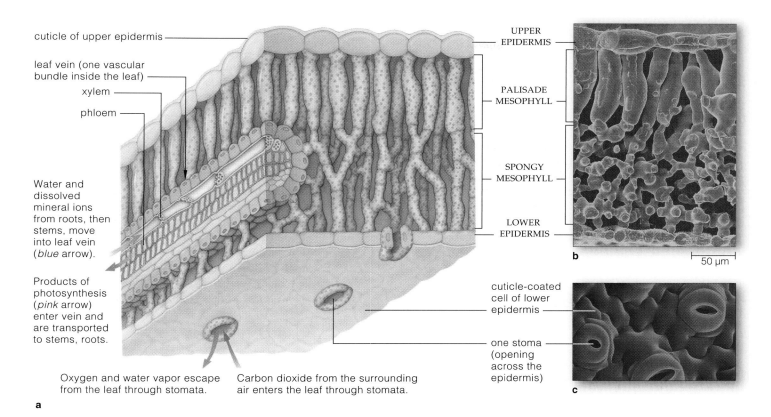

cuticle of upper epidermis

leaf vein (one vascular bundle inside the leaf)

xylem

phloem

Water and dissolved mineral ions from roots, then stems, move into leaf vein (*blue* arrow).

Products of photosynthesis (*pink* arrow) enter vein and are transported to stems, roots.

Oxygen and water vapor escape from the leaf through stomata.

Carbon dioxide from the surrounding air enters the leaf through stomata.

a

UPPER EPIDERMIS

PALISADE MESOPHYLL

SPONGY MESOPHYLL

LOWER EPIDERMIS

b

50 µm

cuticle-coated cell of lower epidermis

one stoma (opening across the epidermis)

c

Figure 29.16 (**a**) Diagram of leaf structure for many kinds of flowering plants. (**b**) Scanning electron micrograph of the tissue organization of a leaf from the kidney bean plant (*Phaseolus*). Notice the compact organization of the epidermal cells. (**c**) Stomata, each a tiny opening across the epidermis, appear when paired guard cells are in their plumped configuration. See also Figure 7.17.

LEAF EPIDERMIS   Epidermis covers every leaf surface exposed to the air. It may be smooth, sticky, or slimy, with "hairs," scales, spikes, hooks, glands, and other surface specialties. The Chapter 30 introduction has two examples. A cuticle covers the sheetlike, compact array of epidermal cells; it restricts the loss of precious water (Figures 29.9 and 29.16). Most leaves have many more stomata on their lower surface. In arid or cold habitats, the stomata are often located at depressions in the leaf surface, together with thickly coated epidermal hairs. Both leaf adaptations help conserve water.

MESOPHYLL—PHOTOSYNTHETIC GROUND TISSUE   As you read earlier, **mesophyll** is a type of parenchyma specializing in photosynthesis. Inside a leaf, most of its cells are exposed to air spaces (Figure 29.16). Carbon dioxide reaches cells by diffusing into the leaf through stomata and through the air inside; oxygen diffuses the opposite way. Adjoining cells exchange substances fast at plasmodesmata. These junctions freely interconnect the cytoplasm of both cells, as Section 4.11 describes.

Leaves oriented perpendicular to the sun have two mesophyll regions. Attached to the upper epidermis is

palisade mesophyll—columnar parenchymal cells with more chloroplasts and more photosynthetic potential, compared to cells of the *spongy* mesophyll layer below them (Figure 29.16). Monocot leaves grow vertically and intercept light from all directions. As you might expect, their mesophyll is not organized as two layers.

VEINS—THE LEAF'S VASCULAR BUNDLES   Leaf **veins** are vascular bundles, usually strengthened with fibers. Their continuous strands of xylem rapidly move water and dissolved nutrients to all mesophyll cells, and the continuous strands of phloem carry the photosynthetic products—especially sugars—away from them. In most dicots, the veins branch lacily into a number of minor veins embedded inside mesophyll. In most monocots, the veins are more or less similar in length, and they run parallel with the leaf's long axis (Figure 29.15).

A leaf's structure is adapted for sunlight interception, gas exchange, and distribution of water, dissolved nutrients, and photosynthetic products. Leaves of each species have a characteristic size, shape, and often surface specializations.

PRIMARY STRUCTURE OF ROOTS

Taproot and Fibrous Root Systems

When a seed germinates, the first part to poke through the seed coat is a **primary root** (Figure 29.17). In nearly all dicot seedlings, it increases in girth while it grows downward. Later on, **lateral roots** form in the primary root's tissues, perpendicular to its axis, and then erupt through epidermis. Youngest lateral roots are closest to the root tip. A **taproot system** is a primary root with its lateral branchings. Dandelions, carrots, and oak trees are examples of plants with this system (Figure 29.17c).

By comparison, the primary root of most monocots, such as grasses, is short-lived. In its place, adventitious roots arise from the stem, and then lateral roots branch from these. (*Adventitious* means the structures form at an unusual location.) The lateral roots are all more or less alike in diameter and length. Together, roots that form this way are a **fibrous root system** (Figure 29.17d).

Internal Structure of Roots

In Figure 29.17a, notice the meristems inside the tip of one root. Many cellular descendants of these meristems divide, enlarge, elongate, and become cells of primary tissue systems. Notice also the root cap, a dome-shaped mass of cells at the tip. The apical meristem produces the cap, which in turn protects the meristem.

Protoderm gives rise to root epidermis, the plant's absorptive interface with the soil. Some epidermal cells send out extensions called **root hairs**. Collectively, root hairs enormously increase the surface area available for taking up water and dissolved nutrients. Only first-time or foolish gardeners would yank a plant from the ground when transplanting it. Such yanking would tear off too much of the highly fragile absorptive surface.

The apical meristem also gives rise to the ground tissue system and to the **vascular cylinder**. A vascular

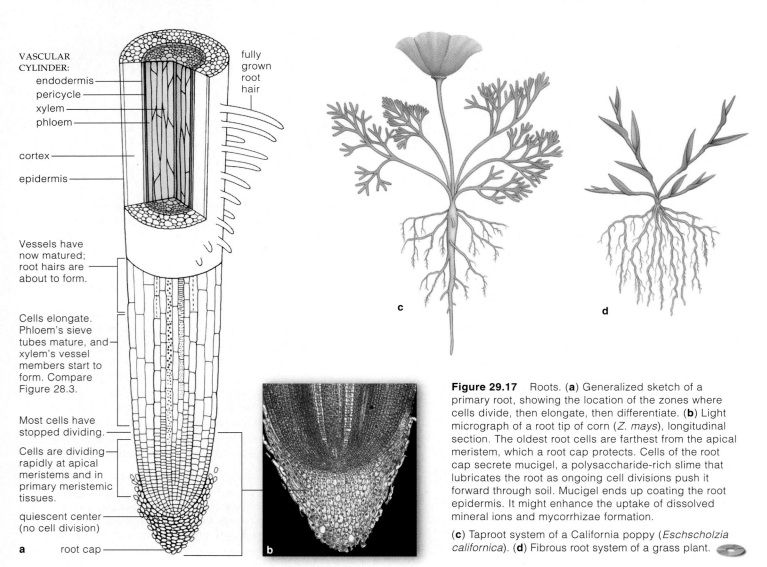

VASCULAR CYLINDER:
endodermis
pericycle
xylem
phloem

cortex

epidermis

fully grown root hair

Vessels have now matured; root hairs are about to form.

Cells elongate. Phloem's sieve tubes mature, and xylem's vessel members start to form. Compare Figure 28.3.

Most cells have stopped dividing.

Cells are dividing rapidly at apical meristems and in primary meristemic tissues.

quiescent center (no cell division)

a   root cap

b

c

d

Figure 29.17 Roots. (**a**) Generalized sketch of a primary root, showing the location of the zones where cells divide, then elongate, then differentiate. (**b**) Light micrograph of a root tip of corn (*Z. mays*), longitudinal section. The oldest root cells are farthest from the apical meristem, which a root cap protects. Cells of the root cap secrete mucigel, a polysaccharide-rich slime that lubricates the root as ongoing cell divisions push it forward through soil. Mucigel ends up coating the root epidermis. It might enhance the uptake of dissolved mineral ions and mycorrhizae formation.

(**c**) Taproot system of a California poppy (*Eschscholzia californica*). (**d**) Fibrous root system of a grass plant.

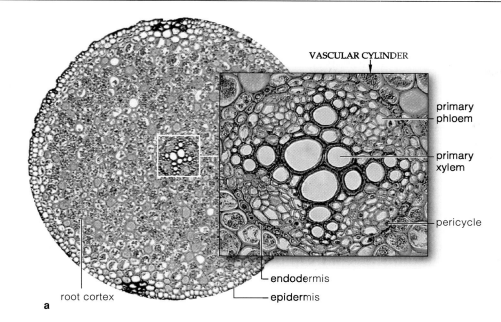

VASCULAR CYLINDER

primary phloem

primary xylem

pericycle

endodermis

epidermis

root cortex

a

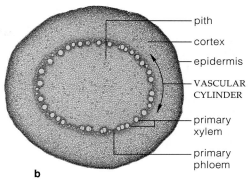

pith

cortex

epidermis

VASCULAR CYLINDER

primary xylem

primary phloem

b

Figure 29.18 (**a**) Young root of a buttercup (*Ranunculus*), transverse section. The inset is a detail of its vascular cylinder. (**b**) Root of corn (*Z. mays*), transverse section. Its vascular cylinder divides the ground tissue into two zones—cortex and pith.

cylinder consists of primary xylem and phloem and one or more layers of cells called the pericycle. Figure 29.18*a* shows a vascular cylinder at the center of a dicot root's cortex. Figure 29.18*b* shows how the vascular cylinder divides the ground tissue system of one type of monocot into cortex and pith regions. With either pattern, there are plenty of air spaces between cells of the ground tissue system, so oxygen can easily diffuse

through them. Like other cells in the plant, all living root cells depend on oxygen for aerobic respiration.

When water enters a root, it moves from cell to cell until it reaches the endodermis, a layer of cells around the vascular cylinder. Where endodermal cells abut, their walls are waterproofed, so incoming water is forced to pass through their cytoplasm. As described in Chapter 30, this arrangement controls the movement of water and dissolved substances into the vascular cylinder.

The pericycle is just inside the endodermis. Some of its cells divide repeatedly and form lateral roots, which erupt through the cortex and epidermis (Figure 29.19).

Regarding the Sidewalk-Buckling, Record-Breaking Root Systems

Unless tree roots start to buckle a sidewalk or choke off a sewer line, most of us do not pay much attention to flowering plant root systems. Roots mine the soil for water and minerals, and most reach a depth of 2 to 5 meters. In hot deserts, where free water is scarce, one hardy mesquite shrub sent its roots down 53.4 meters (175 feet) near a stream bed. Some cacti have shallow roots radiating outward for 15 meters. Someone once measured the roots of a young rye plant that had been growing for four months in 6 liters of soil water. If the surface area of that root system were laid out as one sheet, it would occupy more than 600 square meters!

Roots provide a plant with a tremendous surface area for absorbing water and solutes. Taproot systems consist of a primary root and lateral branchings. Fibrous root systems consist of adventitious roots that replace the primary root.

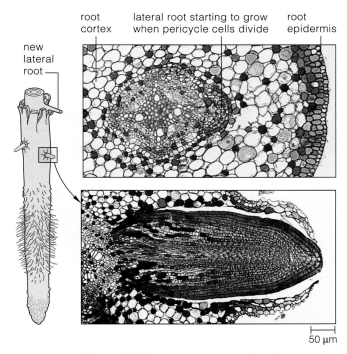

root cortex

lateral root starting to grow when pericycle cells divide

root epidermis

new lateral root

50 µm

Figure 29.19 Lateral root formation in a primary root from a willow tree (*Salix*), transverse section.

ACCUMULATED SECONDARY GROWTH—THE WOODY PLANTS

Flowering plant life cycles extend from germination to seed formation, then death. **Annuals** complete their life cycle in a single growing season, and they typically are "nonwoody," or herbaceous. Examples are marigolds and alfalfa. **Biennials** such as carrots can live for two consecutive growing seasons. Their roots, stems, and leaves form the first season; flowers form, seeds form, and the plant dies the next season. **Perennials** continue vegetative growth and seed formation year after year. And roots and stems thicken in many of them.

Woody and Nonwoody Plants Compared

Like all gymnosperms, some monocots, many eudicots, and magnoliids add secondary growth in two or more growing seasons; they are *woody* plants. Early in life, their stems and roots are like those of nonwoody plants. Differences emerge after their lateral meristems become active and start producing large amounts of secondary vascular tissues, especially the secondary xylem. Again, periderm replaces epidermis on roots and stems that continue to add secondary growth.

The differences are especially pronounced in some of the perennial plants in which the vascular cambium

has become reactivated each growing season, often for hundreds or thousands of years. Ongoing meristematic activity has resulted in giants. To give an example, at last measure, the massive trunk of one coast redwood (*Sequoia sempervirens*) towered more than 110 meters above the forest floor. By one estimate, that redwood's accumulated secondary growth may weigh close to 100 metric tons. Another example: The tree with the largest girth is one of the chestnuts (*Castanea*) that is growing in Sicily. To walk completely around the base of it, you would have to pace off 58 meters.

What Happens at the Vascular Cambium?

Massive stems and roots originate with the vascular cambium. Look at Figure 29.20. Each spring, primary growth resumes at this stem's buds; secondary growth is added *inside* it. Fully formed vascular cambium in a stem is like a cylinder, one or a few cells thick. Some of its cells (fusiform initials) give rise to secondary xylem and phloem that extend *lengthwise* through the stem. Other cells of the vascular cambium (ray initials) give rise to *horizontal* rays of parenchyma, in a pattern a bit like a sliced pie. Through these vascular tissues, water

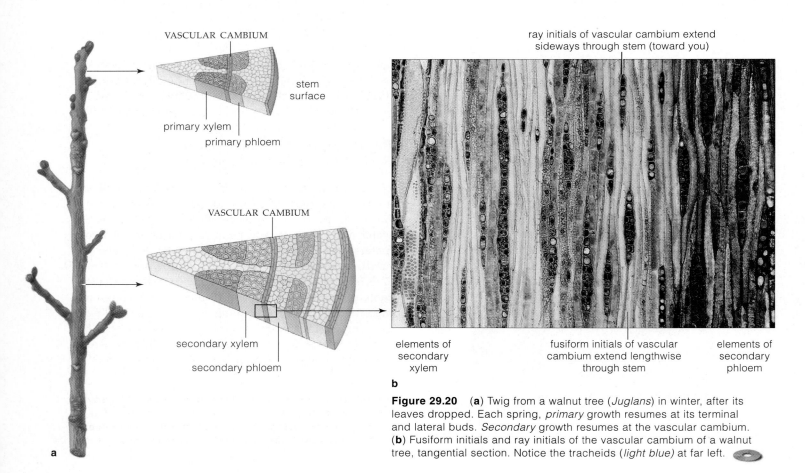

Figure 29.20 (**a**) Twig from a walnut tree (*Juglans*) in winter, after its leaves dropped. Each spring, *primary* growth resumes at its terminal and lateral buds. *Secondary* growth resumes at the vascular cambium. (**b**) Fusiform initials and ray initials of the vascular cambium of a walnut tree, tangential section. Notice the tracheids (*light blue*) at far left.

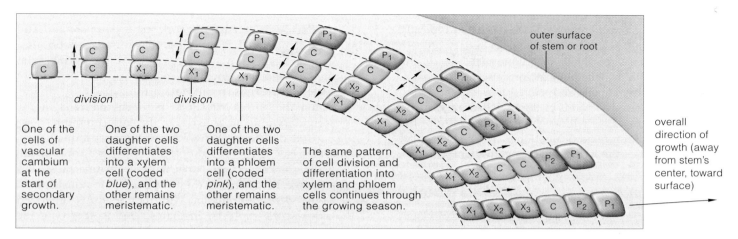

Figure 29.21 Pattern of activity at vascular cambium. Reading left to right, ongoing cell divisions enlarge the inner core of secondary xylem and displace vascular cambium toward the stem or root surface.

Within the figure:

outer surface of stem or root

overall direction of growth (away from stem's center, toward surface)

One of the cells of vascular cambium at the start of secondary growth.

One of the two daughter cells differentiates into a xylem cell (coded *blue*), and the other remains meristematic.

One of the two daughter cells differentiates into a phloem cell (coded *pink*), and the other remains meristematic.

The same pattern of cell division and differentiation into xylem and phloem cells continues through the growing season.

division   division

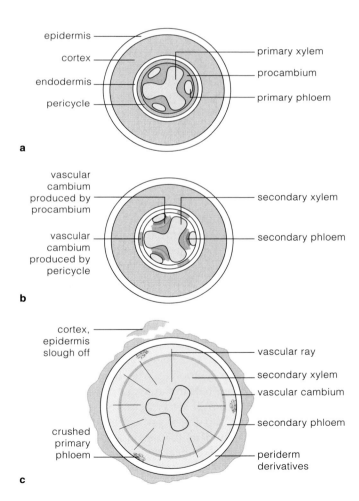

a
- epidermis
- cortex
- endodermis
- pericycle
- primary xylem
- procambium
- primary phloem

b
- vascular cambium produced by procambium
- vascular cambium produced by pericycle
- secondary xylem
- secondary phloem

c
- cortex, epidermis slough off
- crushed primary phloem
- vascular ray
- secondary xylem
- vascular cambium
- secondary phloem
- periderm derivatives

Figure 29.22 Secondary growth in one type of woody root. (**a**) This is how tissues are organized as primary growth ends. (**b,c**) A thin cylinder of vascular cambium forms and gives rise to secondary xylem and phloem. Cell divisions are parallel with the vascular cambium. The cortex ruptures as the root thickens.

---

and solutes travel up, down, and sideways through the enlarging woody stem.

Figure 29.21 shows the growth pattern at vascular cambium. Secondary xylem forms on this meristematic tissue's *inner* face. Secondary phloem forms on its *outer* face. As the inner core of xylem gradually thickens, it displaces meristematic cells toward the surface of the stem. Its cells also maintain a ring of vascular cambium by dividing sideways, in a widening circle.

We have been focusing on how stems thicken, but bear in mind that secondary xylem and phloem also form at vascular cambium in the plant's roots. Figure 29.22 shows one of the patterns of secondary growth at vascular cambium in the root of a typical plant.

What are the selective advantages of woody stems and roots? Remember, like all other organisms, plants compete for resources. Plants having taller stems or broader canopies that defy the pull of gravity intercept more energy streaming in from the sun. With a greater supply of energy for photosynthesis, they have the metabolic means to form large root and shoot systems. With larger systems, they can be more competitive in acquiring resources—and ultimately to be successful, in reproductive terms, in particular habitats.

In woody plants, secondary vascular tissues form at a ring of vascular cambium inside older stems and roots. Wood is an accumulation of secondary xylem especially.

With their sturdier tissues, woody plants defy gravity and grow taller and broader. Where competition for sunlight is intense, the ones that intercept the most sunlight win. Other factors being equal, having more energy to drive photosynthesis provides advantages in terms of metabolic capabilities, growth, and reproductive success.

A CLOSER LOOK AT WOOD AND BARK

Where secondary xylem (wood) is extensive, it typically makes up about 90 percent of a tree. Secondary phloem is restricted to a relatively thin zone just outside the vascular cambium. This phloem consists of thin-walled, living parenchyma cells and sieve tubes that are often interspersed between bands of thick-walled, reinforcing fibers. Only the tubes within about a centimeter of the vascular cambium remain functional. The rest are dead and help protect the living cells beneath them.

Formation of Bark

As seasons pass and a tree ages, its inner core of xylem continues its outward expansion. The resulting pressure is directed toward the stem or root surface. Eventually it ruptures the cortex and the outer part of secondary phloem. Some cortex and epidermis split away. A new surface cover, the periderm, forms from cork cambium. Together, the periderm and secondary phloem constitute **bark**. In other words, bark is composed of all tissues outside the vascular cambium (Figures 29.23 and 29.24).

Periderm consists of cork, new parenchyma, and cork cambium that produces these tissues. Soon after vascular cambium forms, cork cambium forms from the outermost parenchyma cells of the stem or the root cortex. Such cells, recall, retain the capacity to divide. When the cortex ruptures, parenchyma cells in the secondary phloem give rise to the cork cambium.

Cell divisions at the cork cambium produce **cork**. This tissue is composed of densely packed rows of cell walls, thickened with suberin. Only its innermost cells are alive, because only they have access to nourishment from xylem and phloem. With its numerous suberized layers, cork protects, insulates, and waterproofs the stem or root surface. Cork also forms over wounded tissues. When leaves are about to drop, cork forms at the place where petioles attach to stems.

Like all living plant cells, the cells in woody stems and roots require oxygen for aerobic respiration and give off carbon dioxide wastes. So how do these gases get across the suberized, corky surface of bark? They do so through lenticels, which are localized areas where the packing of cork cells is loosened up a bit. Those dark spots you may have noticed on a wine bottle's cork are all that is left of lenticels.

Heartwood and Sapwood

Wood's appearance and function change as a stem or root grows older. The core becomes **heartwood**. This is a dry tissue that is no longer transporting water and solutes. It helps the tree defy gravity and is a dumping ground for various metabolic wastes, including resins, tannins, gums, and oils. Eventually, wastes clog and fill in the oldest xylem pipelines. They commonly darken heartwood, make it more aromatic, and strengthen it.

In the early 1900s, when lumbermen were active in California's groves of old-growth redwoods, someone cut a tunnel through heartwood of a few of the biggest trees, the better to drive an automobile through them.

Sapwood is all of the secondary growth in between the vascular cambium and heartwood (Figure 29.24*a*). Unlike heartwood, sapwood is wet, usually pale, and not as strong. Maple trees are an example. Each spring, from early March through early April, New Englanders insert metal tubes into the sapwood of sugar maples. Sap, a sugar-rich fluid in the secondary xylem, drips through the tubes, into buckets positioned below.

Figure 29.23 The thick, fire-resistant bark of a champion of secondary growth—a coast redwood (*Sequoia sempervirens*).

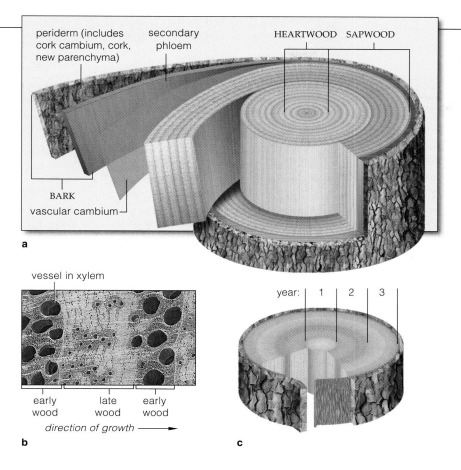

Figure 29.24 (**a**) Stem having extensive secondary growth. (**b**) Scanning electron micrograph of early and late wood cut from red oak (*Quercus rubra*). (**c**) Radial cut through a stem with three annual rings. The first year, the stem put on primary and some secondary growth; the next two years, it added secondary growth.

periderm (includes cork cambium, cork, new parenchyma)

secondary phloem

HEARTWOOD SAPWOOD

BARK

vascular cambium

vessel in xylem

early wood   late wood   early wood

direction of growth ⟶

year: 1 2 3

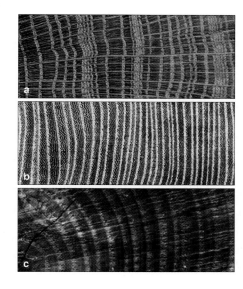

Figure 29.25 Growth layers, or tree rings, of (**a**) oak and (**b**) elm, two hardwood dicots that are durable and strong. (**c**) Pine growth layers. Pine is a softwood. It is lightweight, resists warping, and grows faster than the hardwoods. It is commercially farmed as a source of relatively inexpensive lumber.

Each growth layer shown corresponds to one growing season. (The elm sequence spans the years 1911 to 1950.) Differences in the widths of growth layers correspond to shifts in climate, including water availability. Count the rings and you have clues to a tree's age and to climates and life in the past.

Early Wood, Late Wood, and Tree Rings

Vascular cambium becomes inactive for part of the year in regions having cool winters or prolonged dry spells. *Early* wood, formed at the start of the growing season, has large-diameter, thin-walled xylem cells. By contrast, *late* wood, which forms in dry summers, has xylem cells with smaller diameters and thicker walls. Cut a transverse section from a trunk and you see alternating bands of early and late wood, which reflect the light differently. These visible differences are **growth rings** or, informally, "tree rings" (Figures 29.24 and 29.25).

Seasonal change is predictable in temperate regions, and trees growing there usually add one growth ring per year. In deserts, thunderstorms rumble through at different times of year, and trees respond by adding more than one growth ring in the same season. In the tropics, seasonal change is almost nonexistent. That is why growth rings are not a feature of tropical trees.

Oak, hickory, and other dicot trees that evolved in temperate and tropical regions are all **hardwoods**; they have vessels, tracheids, and fibers in their xylem. Pines, redwoods, and other conifers are **softwood** trees; their xylem contains tracheids and rays of parenchyma, but no vessels or fibers. Lacking fibers, the trees are weaker and less dense than the hardwoods (Figure 29.25).

Limits to Secondary Growth

Some trees, including bristlecone pines and redwoods, gradually put on secondary growth for centuries. But most die far sooner from old age and environmental insults. Compartmentalization, that response to attack you read about in Section 28.4, eventually shuts off the flow of water and solutes through the vascular system.

Bark consists of all living and nonliving tissues outside the vascular cambium—that is, secondary phloem and periderm.

Periderm consists of cork (the outermost covering of woody stems and roots), cork cambium, and new parenchyma.

Wood may be classified by its location and functions (as in heartwood versus sapwood) and by the type of plant (many dicots produce hardwood, and conifers produce softwood).

SUMMARY   **Gold** indicates text section

1. Seed-bearing vascular plants include gymnosperms and angiosperms (the flowering plants). Their shoots (stems, leaves, and other structures) and roots consist of dermal, ground, and vascular tissue systems. *29.1*

2. All plant growth originates at meristems, localized regions of cells that retain the capacity to divide. *29.1*

   a. Primary growth (lengthening of stems and roots) originates at apical meristems in root and shoot tips.

   b. In many plants, secondary growth (increases in diameter) originates inside stems and roots, at lateral meristems called vascular cambium and cork cambium.

3. Parenchyma, sclerenchyma, and collenchyma are the simple tissues, with only one cell type (Table 29.1). *29.2*

   a. Parenchyma cells, alive and metabolically active at maturity, make up the bulk of ground tissue systems. They function in a variety of tasks. The ones making up mesophyll, for example, are photosynthetic.

   b. Collenchyma supports growing plant parts. With its thick, lignified cell walls, sclerenchyma functions in mechanical support.

4. Complex tissues include vascular tissues (xylem and phloem) and dermal tissues (epidermis and periderm). Each has two or more cell types (Table 29.1). *29.2*

   a. Vascular tissues distribute water and dissolved substances throughout a plant. Vascular bundles (each having xylem and phloem clustered together inside a cellular sheath) thread through the ground tissue.

   b. Water-conducting cells of xylem are not alive at maturity. Their lignified, pitted walls interconnect and serve as pipelines for water and dissolved minerals.

   c. Phloem's conducting cells are alive at maturity. The cytoplasm of adjoining cells interconnects across perforated end and side walls. In leaves, sugars and other photosynthetic products are loaded into the cells; often companion cells assist in this. The unloading takes place wherever cells are growing or storing food.

   d. Epidermis covers and protects the outer surfaces of primary plant parts. Periderm replaces epidermis on plants showing extensive secondary growth. *29.2, 29.6*

5. Stems support upright growth and conduct water and solutes through their vascular bundles. Monocot stems often have vascular bundles distributed through ground tissue. Most dicot stems have a ring of bundles dividing the ground tissue into cortex and pith. *29.3*

6. Leaves have veins and mesophyll (photosynthetic parenchyma) between the upper and lower epidermis. Air spaces around the photosynthetic cells enhance gas exchange. Water vapor and gases cross the epidermis through numerous tiny openings called stomata. *29.4*

7. Roots absorb water and dissolved mineral ions for distribution to aboveground parts. Most anchor plants and store food. Some help support shoots. *29.1, 29.5*

Table 29.1 *Summary of Flowering Plant Tissues and Their Components*

SIMPLE TISSUES

Parenchyma	Parenchyma cells
Collenchyma	Collenchyma cells
Sclerenchyma	Fibers or sclereids

COMPLEX TISSUES

Xylem	Conducting cells (tracheids, vessel members); parenchyma cells; sclerenchyma cells
Phloem	Conducting cells (sieve-tube members); parenchyma cells; sclerenchyma cells
Epidermis	Undifferentiated cells; also guard cells and other specialized cells
Periderm	Cork; cork cambium; new parenchyma

8. Wood (secondary xylem) is classified by location and function (as in heartwood or sapwood) and plant type (hardwood of many dicots, softwood of conifers). Bark consists of secondary phloem and periderm. *29.6, 29.7*

Review Questions

1. List some functions of roots and shoots. *29.1*

2. Name and define the basic functions of a flowering plant's three main tissue systems. *29.1*

3. Describe the differences between:
   a. apical, transitional, and lateral meristems *29.1*
   b. parenchyma and sclerenchyma *29.2*
   c. xylem and phloem *29.2*
   d. epidermis and periderm *29.2, 29.7*

4. Which of the two stem sections below is typical of most dicots? Which is typical of most monocots? Label the main tissue regions of both sections. *29.2*

5. In Figure 29.26, is the plant with the yellow flower a dicot or a monocot? What about the plant with the purple flower? *29.2*

Figure 29.26 Flower of (**a**) St. John's wort (*Hypericum*) and (**b**) an iris (*Iris*).

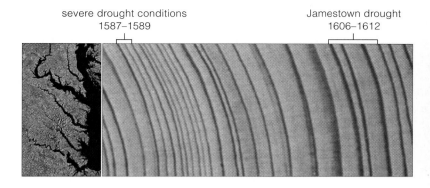

Figure 29.27 *Left:* Map of Virginia's Tidewater region. *Right:* Growth layers of a bald cypress that was living when the first of the English colonizers were in North America.

Self-Quiz ANSWERS IN APPENDIX III

1. Roots and shoots lengthen through activity at _____ .
   a. apical meristems   c. vascular cambium
   b. lateral meristems   d. cork cambium

2. Older roots and stems thicken through activity at _____ .
   a. apical meristems   c. vascular cambium
   b. cork cambium   d. both b and c

3. Soft, moist plant parts consist mostly of _____ cells.
   a. parenchyma   c. collenchyma
   b. sclerenchyma   d. epidermal

4. Xylem and phloem are _____ tissues.
   a. ground   b. vascular   c. dermal   d. both b and c

5. _____ conducts water and ions; _____ conducts food.
   a. Phloem; xylem   c. Xylem; phloem
   b. Cambium; phloem   d. Xylem; cambium

6. Buds give rise to _____ .
   a. leaves   c. stems
   b. flowers   d. all of the above

7. Mesophyll consists of _____ .
   a. waxes and cutin   c. photosynthetic cells
   b. lignified cell walls d. cork but not bark

8. In early wood, cells have _____ diameters, _____ walls.
   a. small; thick   c. large; thick
   b. small; thin   d. large; thin

9. Match the plant parts with the most suitable description.
   ____ apical meristem   a. masses of xylem
   ____ lateral meristem   b. source of primary growth
   ____ xylem, phloem   c. corky surface covering
   ____ periderm   d. source of secondary growth
   ____ vascular cylinder   e. distribution of water, food
   ____ wood   f. central column in roots

Critical Thinking

1. Sylvia lives in Santa Barbara, where droughts are common and a long-term abundance of water is not. She replaced most of her garden with drought-tolerant plants and cut back the size of the lawn. The lawn does not get a light sprinkling every day. Sylvia waters it only twice a week in the evening, after the sun goes down. Then the lawn gets a good soak to a depth of several inches. Why is her strategy good for lawn grasses?

2. *Girdling* means making a continuous cut right through the vertical phloem all the way around a tree trunk. Without the phloem, food from leaves cannot reach roots. If the roots die, so, in time, will the tree. In northern California, anti-logging activists and loggers have been in conflict. Not too long ago, a redwood given the name "Luna" became a symbol for the activists. One night an anti-activist buzz-sawed around most of Luna's trunk. The partially girdled tree hasn't died yet. Do some research and then speculate on why someone wanted to destroy the symbolic tree in the first place.

3. Oscar and Lucinda meet on a trip through a tropical rain forest and fall in love. In the exuberance of the moment, he carves their initials into the bark of a small tree. They never do get together, though. Ten years later, the still-heartbroken Oscar searches for the tree. Given what you know about primary and secondary growth, will he find the carved initials higher relative to ground level? If he goes berserk and cuts down the tree, what kind of growth rings will he see?

4. Environmental conditions apparently worked against the early immigrants from England who tried to settle in North America. Some attempted to establish a colony on Roanoke Island in 1560 or so, in Virginia's Tidewater region. The colony lasted twenty-seven years, then it vanished from the historical record. Historians thought the colony failed because the people were poor planners. However, tree growth layers tell a very different story.

Scientists extracted a wood core from a bald cypress in the region (they didn't harm the tree). Its growth layers revealed that the colonizers were in the wrong place at the wrong time. They were smack in the middle of the worst prolonged drought to hit the mid-Atlantic seaboard over the past eight hundred years. Between the years 1587 and 1589, drought conditions were especially severe. That is precisely when the Roanoke Island colonizers disappeared (Figure 29.27).

Securing and growing food must have been challenging enough. But the colonizers of this region also had to drink brackish water. During the prolonged drought, salts became more concentrated than ever in their water supply. Either the colonizers dispersed elsewhere or the salts poisoned them.

If the secrets locked inside trees intrigue you, do some research into a field of study called dendroclimatology. For example, find out what growth layers reveal about fluctuations in the climate where you live. See if you can correlate it with human events at the time the changes were occurring.

Selected Key Terms

annual 29.6
bark 29.7
biennial 29.6
bud 29.3
collenchyma 29.2
cork 29.7
cork cambium 29.1
cortex 29.3
cuticle (plant) 29.2
dermal tissue system 29.1
epidermis 29.2
eudicot (true dicot) 29.1
fibrous root system 29.5
ground tissue
   system 29.1
growth ring 29.7

hardwood 29.7
heartwood 29.7
lateral root 29.5
leaf 29.4
magnoliid 29.1
meristem 29.1
mesophyll 29.4
monocot 29.1
parenchyma 29.2
perennial 29.6
periderm 29.7
phloem 29.2
pith 29.3
primary root 29.5
root 29.1
root hair 29.5

sapwood 29.7
sclerenchyma 29.2
shoot 29.1
softwood 29.7
stoma
   (stomata) 29.2
taproot system 29.5
vascular bundle 29.3
vascular
   cambium 29.1
vascular
   cylinder 29.5
vascular tissue
   system 29.1
vein (leaf) 29.4
xylem 29.2

Readings

Raven, P., et al. 1999. *Biology of Plants.* Sixth edition. New York: Freeman.

On-Line readings at Student Guide for InfoTrac:
www.brookscole.com/biology

PLANT NUTRITION AND TRANSPORT

Flies for Dinner

How often do we think that plants actually do anything impressive? Being mobile, intelligent, and emotional, we humans tend to be fascinated more with ourselves than with immobile, expressionless plants. Yet plants don't just stand around soaking up sunlight. Consider the Venus flytrap (*Dionaea muscipula*), a flowering plant native to the bogs of North and South Carolina. Its two-lobed, spine-fringed leaves open and close like a steel trap (Figure 30.1*a–d*). Like all other plants, it cannot grow properly without nitrogen and other nutrients, which happen to be scarce in the soil of bogs. However, plenty of insects fly in from places around the bogs.

Sticky sugars ooze from epidermal glands onto the surface of the flytrap's leaf. The sugars entice insects to land. As they do, they brush against hairlike structures that project from the leaf surface. These are triggers for the trap. When an insect touches two hairs at the same time or the same hair twice in rapid succession, the two lobes of the leaf snap shut. Now digestive juices pour out from cells of the leaf. They pool around the insect, dissolve it, and so release nutrients from it. In this way the Venus flytrap makes its own nutrient-rich water, which it proceeds to absorb!

The Venus flytrap is only one of several species of **carnivorous plants**. We call them this even though it takes a leap of the imagination to put their mode of nutrient acquisition—a form of extracellular digestion and absorption—in the same category as the chompings of lions, dogs, and other meat eaters. Besides, not all carnivorous plants actively spring traps. Some types have fluid-filled traps into which prey slip, slide, or fall and then simply drown (Figure 30.2).

All carnivorous plants evolved in habitats where nitrogen and other nutrients are hard to come by. For instance, you may also come across plants with bizarre nutrient-acquiring habits in shallow freshwater lakes and streams, which contain only dilute concentrations of dissolved minerals.

Given the variety and numbers of insects and other animals that attack plants, you can just imagine how endearing the carnivorous plants are to botanists. With their plucky modes of nutrition, these plants also are a fine way to start thinking about **plant physiology**—the study of adaptations by which plants function in their environment. As you already know, nearly all plants are photoautotrophs that use energy from sunlight to drive the synthesis of organic compounds from water, carbon dioxide, and some minerals. Like people, they do not

VENUS FLYTRAP, OPEN FOR DINNER

Figure 30.1 Do plants take nutrition seriously? You bet. (**a**) A Venus flytrap (*Dionaea muscipula*), a carnivorous plant. It makes up for scarce nutrients in its habitat by absorbing them from animals that land on its leaves. (**b**) A fly stuck in sugary goo on a lobed leaf. (**c**) It brushes against hairlike triggers projecting from the leaf; the base of one is shown here. (**d**) An activated leaf snaps shut in just half a second. How? Mesophyll cells right below the epidermis are compressed when the trap is open. Spring the trap and turgor pressure makes the cells decompress abruptly. Whoosh!

base of epidermal hair   epidermal gland

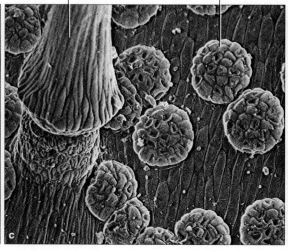

Figure 30.2 Cobra lily (*Darlingtonia californica*). Its leaves form a "pitcher" that is partly filled with digestive juices. Insects lured in by irresistible odors often cannot find the way back out; light shining through the pitcher's patterned dome confuses them. They just wander around and down, adhering to downward-pointing leaf hairs— which are slickened with wax above the potent vat.

have unlimited supplies of the resources necessary to nourish themselves. Of every 1 million molecules of air, only 350 are carbon dioxide. Unlike the soggy habitats of Venus flytraps, most soils are frequently dry. And nowhere except in overfertilized gardens does soil water hold lavish amounts of dissolved minerals. As you continue with these chapters on vascular plants, keep this point in mind: *Many aspects of plant structure and function are responses to low concentrations of vital environmental resources.*

Key Concepts

1. Many aspects of a plant's structure and function are adaptive responses to low concentrations of water, minerals, and other environmental resources.

2. A plant's root system takes up water from soil and also mines the soil for nutrients. For many land plants, mycorrhizae and bacterial symbionts assist in nutrient uptake. In a given habitat, soil properties affect water and nutrient availability for plants.

3. A plant's cuticle and its many stomata function in the conservation of water, a scarce resource in most habitats on land. Stomata are passageways across the epidermis of leaves and, to a lesser extent, stems. When open, stomata permit gas exchange. When closed, they help control water loss.

4. Stomata open during the day, when photosynthesis proceeds. Carbon dioxide diffuses into leaves, oxygen diffuses out—and water loss is rapid. Most plant species conserve water by closing stomata at night.

5. In flowering plants and other vascular plants, the flow of water and solutes through xylem and phloem functionally connects all living cells of roots, stems, and leaves. Xylem serves in the uptake and distribution of water and dissolved mineral ions. Phloem functions in distributing photosynthetically produced sugars and other organic compounds.

6. Water absorbed from soil moves on up through xylem and into leaves. By the process of transpiration, dry air around leaves promotes evaporation through stomata. The force of evaporation pulls continuous columns of water molecules that are hydrogen-bonded to one another from roots to aboveground parts.

7. By the energy-requiring process of translocation, sucrose and other organic compounds are distributed throughout the plant. Organic compounds produced by photosynthetic cells in leaves are loaded into conducting cells of phloem. They are unloaded at the plant's actively growing regions or at storage regions.

PLANT NUTRIENTS AND THEIR AVAILABILITY IN SOILS

Nutrients Required for Plant Growth

So far, we've mentioned nutrients in passing. But exactly what are they? A **nutrient** is any element essential for an organism because no other element can indirectly or directly fulfill its metabolic role. Essential elements for plants are the oxygen, carbon, and hydrogen required for photosynthesis. Plants also need least thirteen other elements (Table 30.1). These usually are dissolved in soil water in ionic forms that reversibly bind with clay. Ions of calcium (Ca^{++}) and potassium (K^+) are examples. Plants easily exchange hydrogen ions for these weakly bound elements.

Nine essential elements are *macro*nutrients. Normally they are required in amounts above 0.5 percent of the plant's dry weight (weighed after all the water has been removed from the plant). The other elements listed are *micro*nutrients; they make up traces (usually a few parts per million) of the dry weight. Even the trace amounts obtained from soil are essential for normal growth.

Properties of Soil

Soil consists of mineral particles mixed with variable amounts of decomposing organic material, or **humus**. Weathering of hard rocks yields these minerals. Dead organisms and organic litter (fallen leaves, feces, and so on) make up the humus. Water and oxygen occupy spaces between the particles and organic bits.

Soils differ in the proportions of mineral particles and how much these are compacted. The three main particle sizes are sand, silt, and clay. Letting beach sand dribble between your fingers can give you a sense that sand particles are large (0.05 to 2 millimeters across). Rubbing silt between your fingers won't tell you much. You won't be able to distinguish among the individual particles, which are only about 0.002 to 0.05 millimeter across. Clay particles are the finest of all.

How suitable is a given soil for plant growth? Is it gummy when wet because it does not have enough air spaces? Does it form hard clods when dry? The answer depends partly on the relative proportions of sand, silt, and clay. The more clay, the finer the soil's texture.

Each clay particle consists of thin, stacked layers of aluminosilicates with negatively charged ions at their surfaces. Clay attracts and binds (adsorbs) positively charged mineral ions dissolved in water that trickles through soil as well as the water molecules themselves. Ions and water cling reversibly to clay. This chemical behavior is vital for all plants. With its high adsorption capacity, clay latches on to many nutrients for plants even as water percolates on past and drains away.

Too much clay is bad for plants. Tightly packed clay particles exclude air spaces, so root cells are deprived

Table 30.1 *Essential Elements and Plant Functioning*

MACRONUTRIENT	Some Functions	Some Deficiency Symptoms	MICRONUTRIENT	Some Functions	Some Deficiency Symptoms
Carbon Hydrogen Oxygen	Raw materials for second stage of photosynthesis	None; all are abundantly available (water, carbon dioxide are sources)	Chlorine	Role in root and shoot growth and photolysis	Wilting; chlorosis; some leaves die
Nitrogen	Protein, nucleic acid, coenzyme, chlorophyll component	Stunted growth; light-green older leaves; older leaves yellow and die (these are symptoms that define a condition called chlorosis)	Iron	Roles in chlorophyll synthesis and in electron transport	Chlorosis; yellow and green striping in leaves of grass species
Potassium	Activation of enzymes; contributes to water–solute balances that influence osmosis*	Reduced growth; curled, mottled, or spotted older leaves; burned leaf edges; weakened plant	Boron	Roles in germination, flowering, fruiting, cell division, nitrogen metabolism	Terminal buds, lateral branches die; leaves thicken, curl, become brittle
Calcium	Component in control of many cell functions; cementing of cell walls	Terminal buds wither, die; deformed leaves; poor root growth	Manganese	Chlorophyll synthesis; coenzyme action	Dark veins, but leaves whiten and fall off
Magnesium	Chlorophyll component; activation of enzymes	Chlorosis; drooped leaves	Zinc	Role in forming auxin, chloroplasts, starch; enzyme component	Chlorosis; mottled or bronzed leaves; root abnormalities
Phosphorus	Phospholipid, nucleic acid, ATP component	Purplish veins; stunted growth; fewer seeds, fruits	Copper	Component of several enzymes	Chlorosis; dead spots in leaves; stunted growth
Sulfur	Component of most proteins, two vitamins	Light-green or yellowed leaves; reduced growth	Molybdenum	Part of enzyme used in nitrogen metabolism	Pale green, rolled or cupped leaves

* All mineral elements contribute to water–solute balances; potassium is notable because there is so much of it.

Figure 30.3 (a) Some of the soil horizons that developed in one habitat in Africa. (b) Profile of a heavily leached soil. Such soils are common in cool, moist coniferous forests. Breakdown of pine needle litter makes the soil water highly acidic, so nutrients are easily leached from surface layers. Acid-resistant materials such as quartz remain and give the layers closest to the surface an ash-gray appearance. Iron and aluminum oxides stain deeper layers.

(c) Erosion forms gullies that channel runoff from the land. As gullies deepen and widen, erosion becomes more rapid. When topsoil is depleted, productivity declines, and fertilizers typically are trucked in to replace lost nutrients.

O HORIZON
Fallen leaves and other organic material littering the surface of mineral soil

A HORIZON
Topsoil, which contains some percentage of decomposed organic material and which is variably deep (only a few centimeters deep in deserts, but elsewhere extending as far as thirty centimeters below the soil surface)

B HORIZON
Compared with the A horizon, larger soil particles, not much organic material, but greater accumulation of minerals; extends thirty to sixty centimeters below soil surface

C HORIZON
No organic material, but partially weathered fragments and grains of rock from which soil forms; extends to underlying bedrock

BEDROCK

of oxygen for aerobic respiration. Packing also retards water penetration into the soil. Runoff, and nutrient loss, is severe in heavy clay soils. The best soils are **loams**, which have roughly the same proportions of sand, silt, and clay.

Humus in soil also affects plant growth. Generally, humus has abundant negatively charged organic acids. It weakly binds and retains dissolved mineral ions of opposite charge. Humus also has a high capacity to absorb and swell with water, then shrink as water is gradually released. Its alternating swelling and shrinking aerates the soil. As decomposers gradually work it over, humus releases nutrients that plants can take up.

In general, the soils that contain 10 to 20 percent humus are most favorable for plant growth. The worst soils of all are less than 10 percent humus or more than 90 percent humus, the latter being a feature of swamps and bogs.

Soils are classified by *profile* properties, their layered characteristics. Soils in different places are in different stages of development. Figure 30.3 has two examples. **Topsoil**, the uppermost soil layer, is the A horizon. This is the most essential layer for plant growth, and its depth is variable from one habitat to the next.

Leaching and Erosion

Leaching is the removal of nutrients from soil as water percolates through it. It is heaviest in sandy soils, which are not as good as clay at binding nutrients. **Erosion** is a movement of land under the force of wind, running water, and ice (Figure 30.3*b,c*). For example, erosion from all the farmlands drained by the Mississippi River puts about 25 billion metric tons of topsoil into the Gulf

of Mexico each year. Whether by leaching or erosion, the loss of nutrients from soil is bad for plants and for all organisms that depend on plants for survival.

Nutrients are essential elements. No other element can substitute for their direct or indirect roles in the metabolic activities that sustain growth and keep organisms alive.

The mineral component of soil includes particles ranging from large-grained sand to silt and fine-grained clay. These particles, clay especially, reversibly bind water molecules and dissolved mineral ions and thereby make them more accessible for uptake by plant roots.

Soil also contains humus, which is a reservoir of organic material, rich in organic acids, in different stages of decay. Most plants grow best in soils having equal proportions of sand, silt, and clay, as well as 10 to 20 percent humus.

HOW DO ROOTS ABSORB WATER AND MINERAL IONS?

In terms of energy outlays, mining the soil for mineral ions and water molecules clinging to clay particles is an expensive task. Plants spend considerable energy on building extensive root systems. Wherever the soil's texture and composition change, new roots must form to replace old ones and infiltrate different regions. It isn't that roots "explore" soil for resources. Rather, the patches of soil where the concentrations of water and mineral ions are greater stimulate outward growth.

Absorption Routes

Think back on the preceding chapter's discussion of a typical root's structure (Section 29.5). Water molecules in soil are only weakly bound to clay particles, so they readily cross the root epidermis and continue into the **vascular cylinder**, a column of vascular tissue. There, a cylindrical layer of cells, an **endodermis**, wraps around the column. A band of waxy deposits—the **Casparian strip**—is embedded in abutting endodermal cell walls (Figure 30.4). Water molecules can't penetrate it. They infiltrate only the unwaxed wall regions, pass through cells, then cross unwaxed wall regions on the opposite side. This is the only way water and solutes move into the vascular cylinder. Like all cells, endodermal cells have many transport proteins embedded in the plasma membrane. The proteins let some solutes but not others cross it (Section 5.7). *The transport proteins of endodermal cells are control points where plants adjust the quantity and types of solutes absorbed from soil water.*

Roots of many plants also have an **exodermis**, a cell layer just beneath their surface (Figure 30.4a). Walls of exodermal cells commonly have a Casparian strip that functions like the one next to a root vascular cylinder.

Specialized Absorptive Structures

ROOT HAIRS   Vascular plants require great amounts of water. Roots of a mature corn plant absorb more than three liters of water daily. They could not do so without **root hairs**. Recall, from the preceding chapter, that root hairs are slender extensions of specialized epidermal cells. They greatly increase the surface area available for absorption (Section 29.5 and Figure 30.5). When a plant is putting on primary growth, its system of roots may develop millions or billions of root hairs.

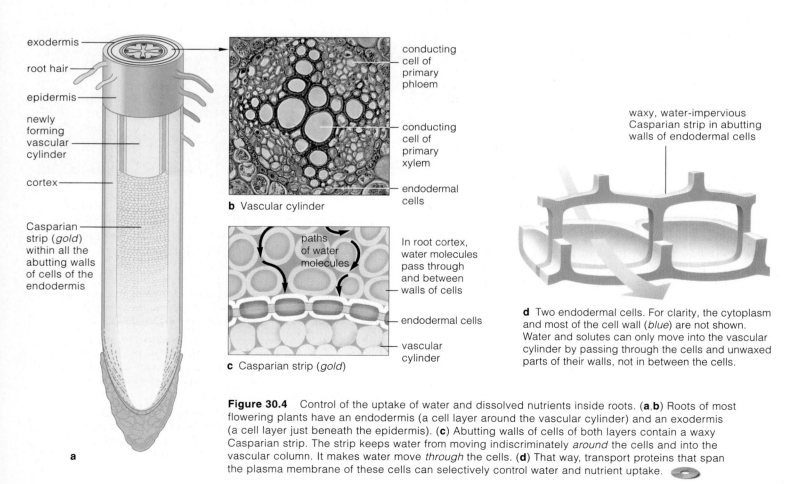

exodermis

root hair

epidermis

newly forming vascular cylinder

cortex

Casparian strip (*gold*) within all the abutting walls of cells of the endodermis

a

conducting cell of primary phloem

conducting cell of primary xylem

endodermal cells

b Vascular cylinder

paths of water molecules

In root cortex, water molecules pass through and between walls of cells

endodermal cells

vascular cylinder

c Casparian strip (*gold*)

waxy, water-impervious Casparian strip in abutting walls of endodermal cells

d Two endodermal cells. For clarity, the cytoplasm and most of the cell wall (*blue*) are not shown. Water and solutes can only move into the vascular cylinder by passing through the cells and unwaxed parts of their walls, not in between the cells.

Figure 30.4   Control of the uptake of water and dissolved nutrients inside roots. (**a,b**) Roots of most flowering plants have an endodermis (a cell layer around the vascular cylinder) and an exodermis (a cell layer just beneath the epidermis). (**c**) Abutting walls of cells of both layers contain a waxy Casparian strip. The strip keeps water from moving indiscriminately *around* the cells and into the vascular column. It makes water move *through* the cells. (**d**) That way, transport proteins that span the plasma membrane of these cells can selectively control water and nutrient uptake.

Figure 30.5 An example of root hairs. These thin extensions of the young root's epidermal cells specialize in absorption of both water and dissolved ions.

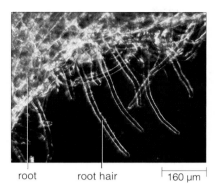

root   root hair   160 μm

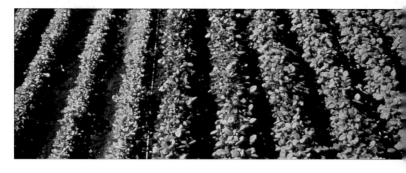

Figure 30.7 Demonstration of the effect of root nodules on plant growth. At *left*, rows of soybean plants growing in nitrogen-poor soil. At *right*, growing in the same soil, plants in these rows were inoculated with *Rhizobium* cells and developed root nodules.

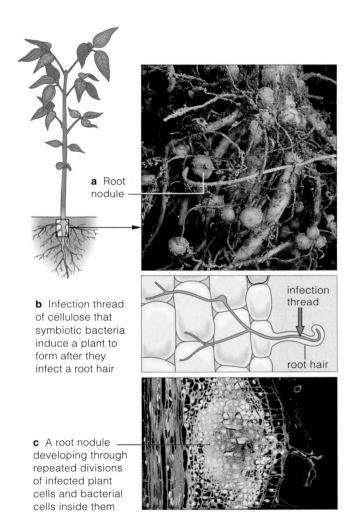

a Root nodule

b Infection thread of cellulose that symbiotic bacteria induce a plant to form after they infect a root hair

infection thread

root hair

c A root nodule developing through repeated divisions of infected plant cells and bacterial cells inside them

Figure 30.6 (**a**) Nutrient uptake at root nodules of legumes that are mutualists with nitrogen-fixing bacteria (*Rhizobium* and *Bradyrhizobium*). When infected by the bacteria, root hair cells form a thread of cellulose deposits. Bacteria use the thread as a highway to invade plant cells in the root cortex.

(**b,c**) Infected plant cells and bacterial cells inside them divide repeatedly. Together they form a swollen mass that becomes a root nodule. The bacteria start fixing nitrogen when membranes of plant cells surround them. The plant takes up some of the nitrogen; the bacteria take up some photosynthetic compounds.

ROOT NODULES Certain bacteria and fungi help many plants absorb dissolved nutrients and get something in return. **Mutualism** is the name for such a two-way flow of benefits between species (Chapter 24). Think about how nitrogen deficiency limits growth of plants. Gaseous nitrogen (N$\equiv$N, or N$_2$) is plentiful in air. But plants don't have the metabolic machinery for **nitrogen fixation**. By this process, enzymes split all N$_2$ covalent bonds and attach both atoms to organic compounds. To obtain high crop yields, farmers apply nitrogen-rich fertilizers or encourage the growth of nitrogen-fixing bacteria in the soil. These bacteria convert N$_2$ to forms that they can utilize. In this case, string beans, peas, alfalfa, clover, and other legumes have an advantage. Nitrogen-fixing bacteria are symbionts in their roots, in localized swellings called **root nodules** (Figures 30.6 and 30.7). Bacterial cells withdraw some of the organic compounds, originally produced by photosynthesis in the leaves. In return, the plants absorb some nitrogen that the bacterial cells secured.

MYCORRHIZAE Also think back on the **mycorrhizae** (singular, mycorrhiza). As described in Chapter 24, a mycorrhiza is a symbiotic interaction between a young root and a fungus. Hence the name, meaning "fungus-root." Fungal filaments—hyphae—form a velvety cover around roots or penetrate the root cells. Collectively, hyphae have a large surface area that absorbs mineral ions from a larger volume of soil than the roots can do. The fungus can absorb sugars and nitrogen-containing compounds from root cells. The root cells obtain some scarce minerals that the fungus is better able to absorb.

Gymnosperms and flowering plant roots control the uptake of water and dissolved nutrients at the vascular cylinder's endodermis and at a similar layer near the root surface.

Root hairs, root nodules, and mycorrhizae greatly enhance the uptake of water and dissolved nutrients.

HOW IS WATER TRANSPORTED THROUGH PLANTS?

Transpiration Defined

By now, you have a sense of how the distribution of water and dissolved mineral ions to all living cells is central to plant growth and functioning. Let's turn now to a model for how water actually moves from a plant's roots to its stems, then into leaves.

Plants, recall, use only a fraction of the water they absorb for growth and metabolism. Most of that water is lost, mainly through the numerous stomata in leaves. Evaporation of water molecules from leaves, stems, and other plant parts is a process called **transpiration**.

Cohesion–Tension Theory of Water Transport

This brings up an interesting question. Assuming that plants lose most of the absorbed water from their leaves, how does water actually get *to* the leaves? What gets it to the uppermost leaves of plants, including redwoods and other trees that may be more than 100 meters tall?

In a plant's vascular tissues, water moves through a complex tissue called **xylem**. Recall, from Section 29.2, that the water-conducting cells of xylem are **tracheids** and **vessel members**. Figure 30.8 provides examples. These cells are dead at maturity, and only their lignin-impregnated walls are left. This means the conducting cells in xylem can't be actively pulling water "uphill."

Some time ago, the botanist Henry Dixon came up with a useful way to explain water transport in plants. By his **cohesion–tension theory**, the water inside xylem is pulled upward by air's drying power, which creates continuous negative pressures—that is, tensions. These tensions extend all the way from leaves to roots. Figure 30.9 illustrates Dixon's theory. Think about these three points as you review the illustration:

First, air's drying power causes transpiration: the evaporation of water from all plant parts exposed to air—but most notably at stomata. Transpiration puts the water confined in xylem's waterproofed conducting tubes into a state of tension. And that tension extends from veins inside leaves, down through the stems, and on into young roots where water is being absorbed.

Second, the unbroken, fluid columns of water show *cohesion*; they resist rupturing while they are pulled up under *tension*. (Here you may wish to reflect on Section 2.5.) The collective strength of all the hydrogen bonds between water molecules in the narrow, tubular xylem cells imparts this cohesion.

Third, for as long as molecules of water continue to escape from a plant, the continuous tension inside the

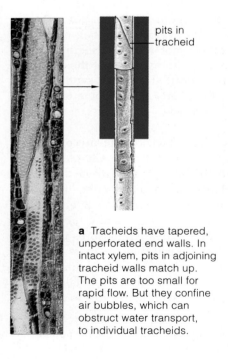

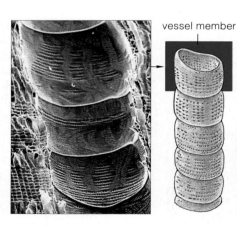

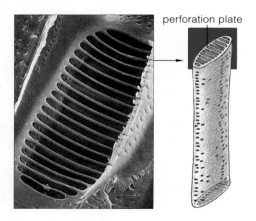

a Tracheids have tapered, unperforated end walls. In intact xylem, pits in adjoining tracheid walls match up. The pits are too small for rapid flow. But they confine air bubbles, which can obstruct water transport, to individual tracheids.

b Close-up of three of the adjoining members that make up a vessel. The thick, finely perforated walls of these dead cells connect one after another to form vessels, another type of water-conducting tube in xylem. The walls of all these tubes contain an abundance of lignin, which strengthens them and also makes them waterproof.

c Perforation plate at the end wall of one type of vessel member. The perforated ends permit water and air bubbles to flow unimpeded through the conducting tube. This might be one reason why natural selection favored the retention of tracheids and vessel members in the same plants.

Figure 30.8 Scanning electron micrographs and sketches of a few types of tracheids and vessel members from xylem. These water-conducting tubes are made of the walls of cells, which are dead at maturity. The cell walls still remain interconnected, and so form the tubes. Tracheids probably evolved before vessel members. Both occur in nearly all vascular plants.

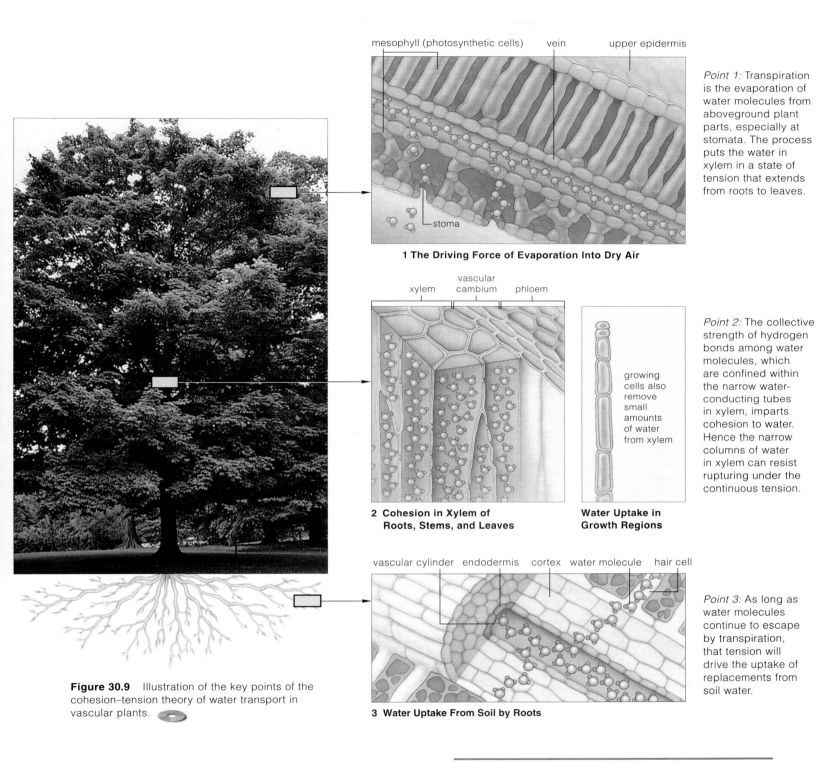

mesophyll (photosynthetic cells)　　vein　　upper epidermis

stoma

1 The Driving Force of Evaporation Into Dry Air

Point 1: Transpiration is the evaporation of water molecules from aboveground plant parts, especially at stomata. The process puts the water in xylem in a state of tension that extends from roots to leaves.

xylem　　vascular cambium　　phloem

growing cells also remove small amounts of water from xylem

2 Cohesion in Xylem of Roots, Stems, and Leaves

Water Uptake in Growth Regions

Point 2: The collective strength of hydrogen bonds among water molecules, which are confined within the narrow water-conducting tubes in xylem, imparts cohesion to water. Hence the narrow columns of water in xylem can resist rupturing under the continuous tension.

vascular cylinder　　endodermis　　cortex　　water molecule　　hair cell

3 Water Uptake From Soil by Roots

Point 3: As long as water molecules continue to escape by transpiration, that tension will drive the uptake of replacements from soil water.

Figure 30.9 Illustration of the key points of the cohesion–tension theory of water transport in vascular plants.

xylem permits more molecules to be pulled upward from the roots, and therefore to replace them.

Hydrogen bonds are strong enough to hold water molecules together inside the xylem. However, they are not strong enough to prevent the water molecules from breaking away from one another during transpiration and then escaping from leaves, through stomata.

Transpiration is the process of evaporation from plant parts.

By the cohesion–tension theory of water transport, this process is the key source of tensions in water in xylem. The tensions extend from leaves to roots, and they allow columns of water molecules that are hydrogen-bonded to one another to be pulled upward through the plant body.

30.4

HOW DO STEMS AND LEAVES CONSERVE WATER?

At least 90 percent of the water transported from roots to a leaf goes right out by evaporating into the air. Cells use only about 2 percent for photosynthesis, membrane functions, and other activities, but that amount must be maintained. So how do plants know how much water they have? They sense **turgor pressure**—the pressure against a cell wall that arises from the movement of water into that cell. When a plant's soft parts are erect, you know that as much water is moving into its cells as is moving out. When the cells lose water, the plant wilts; its soft parts droop and water-dependent events are disrupted (Sections 5.7 and 28.2). Yet plants are not entirely at the mercy of changes in water availability. They have a cuticle, and they have stomata.

The Water-Conserving Cuticle

Even mildly water-stressed plants would rapidly wilt and die without their **cuticle** (Figure 30.10). Epidermal cells secrete this translucent, water-impermeable layer, which coats cell wall regions exposed to the air. At the cuticle surface are deposits of waxes, which are water-insoluble lipids having long fatty-acid tails. The cuticle itself consists of waxes embedded in **cutin**, an insoluble

lipid polymer. Beneath all the waxes, cellulose fibers weave through the cutin. A layer of polysaccharides (pectins) often helps bind the cuticle to the cell walls.

A cuticle does not bar the passage of light rays into photosynthetic parts of the plant. It does restrict water loss. It also restricts the *inward* diffusion of the carbon dioxide necessary for photosynthesis and the *outward* diffusion of oxygen by-products. Recall, from Section 7.7, that a buildup of oxygen in the air spaces inside a leaf has bad effects on the rate of photosynthesis.

Controlled Water Loss at Stomata

How do carbon dioxide and oxygen get past the cuticle-covered, water-conserving epidermis? They cross it in controlled fashion at **stomata** (singular, stoma, a Greek word meaning mouth). Gas exchange and evaporation of water occur mainly at stomata.

Stomata usually are open in the day to help support photosynthesis. Water is lost, but given sufficient soil moisture, roots replace it. At night, stomata are closed. Water is conserved, and carbon dioxide accumulates in leaves as cells engage in aerobic respiration. Like you, plants depend on this energy-releasing pathway to get enough ATP to drive most of their metabolic reactions.

Stomata close mainly in response to water loss. Each stoma is defined by a pair of specialized parenchyma cells called **guard cells** (Figure 30.11). When the pair swell with water, the turgor pressure makes both bend, so a gap (stoma) forms between them. When they lose water and turgor pressure drops, the two cells collapse against each other, so the gap closes.

When a plant is water stressed, its stomata close in response to a signal from the hormone called abscisic acid (ABA). Remember the Section 28.5 introduction to signal reception, transduction, and response? ABA is such a signal. It binds to receptors on each guard cell's plasma membrane. Some experiments indicate that the binding causes the opening of gated channels across the membrane. Calcium ions (Ca^{++}) flow into the cells, where they may act as a second messenger. They cause other channels to open that let chloride ions (Cl^-) and a negatively charged organic compound, malate, flow rapidly from the cytoplasm to the extracellular matrix.

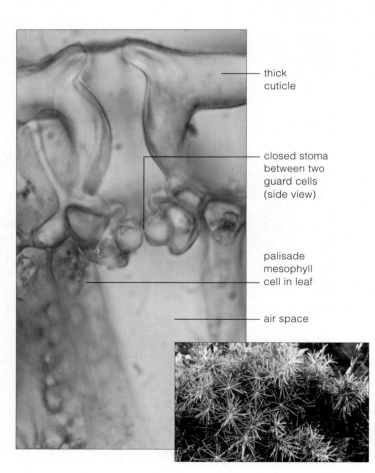

thick cuticle

closed stoma between two guard cells (side view)

palisade mesophyll cell in leaf

air space

Figure 30.10 From Australia, cuticle of the circular, needle-like leaves of *Hakea gibbosa*, known as the rock needlebush, longitudinal section. Water, carbon dioxide, and oxygen cross the cuticle at stomata. Like other plants in seasonally dry habitats, it has a very thick cuticle that helps restrict water loss. (Aquatic plants have thin cuticles or none at all.)

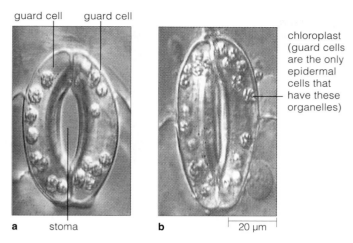

guard cell   guard cell

chloroplast (guard cells are the only epidermal cells that have these organelles)

a stoma   **b**   20 μm

Figure 30.11 Stomatal action. Whether a stoma is open or closed at any given time depends on the shape of two guard cells that define this small opening across a cuticle-covered leaf epidermis. (**a**) This stoma is open. Water entered collapsed guard cells, which swelled under turgor pressure and moved apart, thus forming the stoma. (**b**) This stoma is closed. Water left the swollen guard cells, which collapsed against each other and closed the stoma.

Figure 30.13 (**a**) Stomata at holly leaf surface, set against a backdrop of smog in Central Europe. (**b**) What they look like when a holly plant is growing in industrialized regions. Gritty airborne pollutants clog its stomata and prevent much of the sun's rays from reaching photosynthetic cells in the leaf.

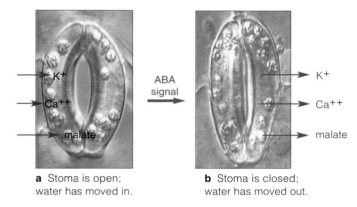

K⁺   ABA signal   K⁺

Ca⁺⁺   Ca⁺⁺

malate   malate

a Stoma is open; water has moved in.   **b** Stoma is closed; water has moved out.

Figure 30.12 Hormonal control of stomatal closure. (**a**) When a stoma is open, high solute concentrations in the cytoplasm of both guard cells have raised the turgor pressure, keeping the cells plumped up. (**b**) In a water-stressed plant, the hormone abscisic acid binds to receptors on the guard cell plasma membrane. It activates a signal transduction pathway that lowers solute concentrations inside the cells, which lowers the turgor pressure and closes the stoma.

The flow creates an electric and concentration gradient for potassium ions (K^+) across the membrane.

With the loss of all these ions, the cytoplasm in the guard cells has a higher water potential (Section 5.7). Now water follows its gradient and moves out of the cells. The turgor pressure falls, and so the stoma closes. Experiments indicate that the stoma opens when ions move in reverse, into the guard cells (Figure 30.12).

Environmental signals also affect whether stomata are opened or closed. They include the concentration of carbon dioxide inside leaves, the incoming light, and

temperature. For example, photosynthesis starts after the sun comes up. As the morning progresses, carbon dioxide levels fall in photosynthetic cells, including the guard cells (Figure 30.12*a*). The decline helps trigger the active transport of K^+ into these cells. So do blue wavelengths of light, which penetrate the atmosphere better as the sun arcs higher in the sky. Water follows potassium ions into the cells. The inward movement increases the turgor pressure and the stoma opens.

Carbon dioxide levels in the cells rise when the sun sets and photosynthetic activity ceases. Potassium, then water, move out of guard cells. And so stomata close.

CAM plants, including most cacti, conserve water differently. They open stomata at night, when they fix carbon dioxide by the metabolic C4 pathway described in Section 7.7. The next day, when their stomata close, CAM plants use carbon dioxide in photosynthesis.

As this section makes clear, plant survival depends on stomatal function. Think about it when you are out and about on smog-shrouded days (Figure 30.13).

Water-dependent events in plants are severely disrupted when water loss exceeds uptake at roots for extended periods. Wilting is one observable outcome.

Transpiration and gas exchange occur mainly at stomata. These numerous small openings span the waxy cuticle, which covers all plant epidermal surfaces exposed to air.

Plants open and close stomata at different times to control water loss, carbon dioxide uptake, and oxygen disposal, all of which affect rates of photosynthesis and plant growth.

HOW ARE ORGANIC COMPOUNDS DISTRIBUTED THROUGH PLANTS?

Whereas xylem distributes water and minerals through the plant, the vascular tissue called **phloem** distributes organic products of photosynthesis. Like xylem, phloem consists of many conducting tubes, fibers, and strands of parenchyma cells. Remember Sections 29.3 and 29.6? They showed phloem's distribution patterns in dicots and monocots. Unlike xylem, phloem has **sieve tubes** through which organic compounds flow swiftly. These tubes consist of *living* cells called sieve-tube members. The cells are positioned side by side and end to end in vascular tissues. Their abutting end walls, called sieve plates, have many pores (Figure 30.14*a*).

Adjoining the long sieve tubes are **companion cells** (Figure 30.14*b*). Like the sieve-tube members, these are differentiated parenchyma cells that help load organic compounds into sieve tubes.

Let's see what happens to sucrose and other organic products of photosynthesis in phloem. Leaf cells use some of the products for their own activities. The rest travels to roots, stems, buds, flowers, and fruits. In most cells, carbohydrates are stored as starch in plastids. Proteins and fats, built from carbohydrates and amino acids distributed to living cells, often are stored in seeds. Avocados and some other fruits also accumulate fats.

Starch molecules are too large for transport across the plasma membrane of the cells that make them and are too insoluble for transport to other regions of the

Figure 30.15 A honeydew droplet that is exuding from the end of an aphid gut. The tubular mouthpart of this small insect penetrated a conducting tube in phloem. The sugary fluid was under high pressure in phloem and was forced out through the gut's terminal opening.

plant. Overall, proteins are too large and fats are too insoluble for transport from storage sites. *But the cells can convert storage forms of organic compounds to smaller solutes that are more easily transported by the phloem.* For instance, the cells degrade starch to glucose monomers. When one of these monomers combines with fructose, the result is sucrose—an easily transportable sugar.

Simple experiments with aphids show that sucrose is the main carbohydrate transported inside phloem. These insects were anesthetized by exposure to high levels of carbon dioxide as they were feeding on juices in conducting tubes of phloem. The body of the aphids was detached from their mouthparts—which were still embedded in the sieve tubes. For most of the plants studied, sucrose was the most abundant carbohydrate in the fluid that was being forced out of the tubes.

Translocation

Translocation is the technical name for the transport of sucrose and all other organic compounds through the phloem of a vascular plant. High pressure drives this process. Often the pressure is five times higher than in automobile tires. Aphids demonstrate the magnitude of it when they force their mouthparts into sieve tubes and feed upon the dissolved sugars. The high pressure can force fluid through an aphid gut and out the other end, as "honeydew" (Figure 30.15). Park a car under some trees being attacked by aphids, and it might get spattered by sticky honeydew droplets, thanks to the high fluid pressure in phloem.

Pressure Flow Theory

Phloem translocates the photosynthetic products along decreasing pressure and solute concentration gradients. The *source* of the flow is any region of the plant where organic compounds are being loaded into sieve tubes. Common sources are mesophylls—the photosynthetic tissues in leaves. The flow ends at a *sink*, which is any plant region where the products are being unloaded, used, or stored. Example: While flowers and fruits are growing during development, they are sink regions.

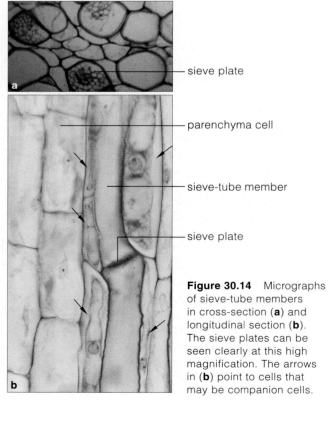

sieve plate

parenchyma cell

sieve-tube member

sieve plate

Figure 30.14 Micrographs of sieve-tube members in cross-section (**a**) and longitudinal section (**b**). The sieve plates can be seen clearly at this high magnification. The arrows in (**b**) point to cells that may be companion cells.

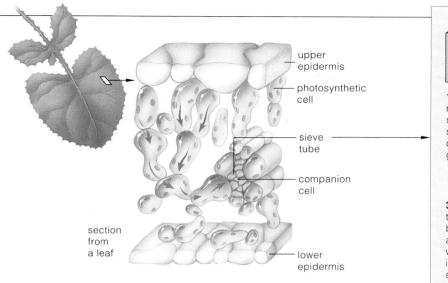

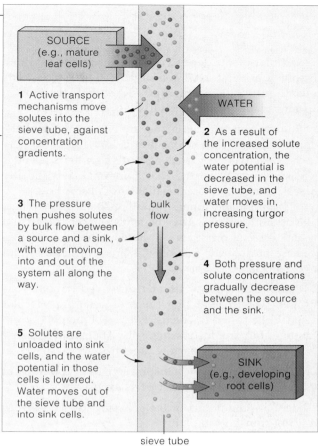

1 Active transport mechanisms move solutes into the sieve tube, against concentration gradients.

2 As a result of the increased solute concentration, the water potential is decreased in the sieve tube, and water moves in, increasing turgor pressure.

3 The pressure then pushes solutes by bulk flow between a source and a sink, with water moving into and out of the system all along the way.

4 Both pressure and solute concentrations gradually decrease between the source and the sink.

5 Solutes are unloaded into sink cells, and the water potential in those cells is lowered. Water moves out of the sieve tube and into sink cells.

sieve tube

a *Loading at a source*. Photosynthetic cells in leaves are a common source of organic compounds that must be distributed through a plant. Small, soluble forms of these compounds move from the cells into phloem (a leaf vein).

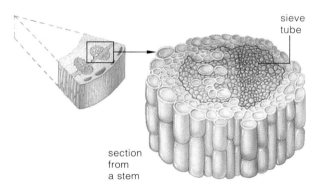

b *Translocation along a distribution path*. Fluid pressure is greatest inside sieve tubes at the source. It pushes the solute-rich fluid to a sink, which is any region where cells are growing or storing food. There, the pressure is lower because cells are withdrawing solutes from the tubes.

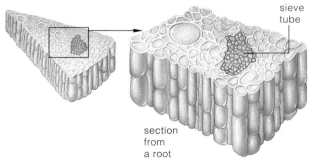

c *Unloading at the sink*. Solutes are unloaded from sieve tubes into cells at the sink; water follows. Translocation continues as long as solute concentration gradients and a pressure gradient exist between the source and the sink.

Figure 30.16 Translocation of organic compounds in sow thistle (*Sonchus*). Review Section 7.6 to get an idea of how translocation relates to photosynthesis in vascular plants.

Why do organic compounds flow from a source to a sink? According to the **pressure flow theory**, internal pressure builds up at the source end of the sieve tube system and *pushes* the solute-rich solution on toward any sink, where solutes are being removed.

Experimental evidence for the pressure flow theory was obtained from studies of sow thistle (*Sonchus*). Use Figure 30.16 to track what happens after sucrose moves from photosynthetic cells into small veins in the leaves of this plant. Companion cells in veins spend energy to load sucrose into adjacent sieve-tube members. As the sucrose concentration increases in the tubes, water also moves in, by osmosis. More internal fluid pressure is exerted against the sieve-tube walls. When the pressure increases, it pushes the sucrose-laden fluid out of the leaf, into the stem, and on to the sink.

Plants store organic compounds in the form of starch, fats, and proteins. They convert the storage forms to sucrose and other small units that are soluble and easily translocated.

Translocation is the distribution of organic compounds to different plant regions. It depends on concentration and pressure gradients in the sieve-tube system of phloem.

Gradients exist as long as companion cells load compounds into the sieve tubes at sources, such as mature leaves, and as long as compounds are removed at sinks, such as roots.

SUMMARY

Gold indicates text section

1. Vascular plants depend on the distribution of water, dissolved mineral ions, and organic compounds to its cells. Figure 30.17 summarizes how that distribution sustains plant growth and survival. *CI, 30.1–30.3*

2. Root systems efficiently take up water and nutrients, which commonly are scarce in soil. *30.1, 30.2*

    a. Collectively, a vascular plant's root hairs (slender extensions of specialized root epidermal cells) greatly increase the surface area available for absorption.

    b. Bacterial and fungal symbionts help many plants take up mineral ions. They benefit from the interaction by using some products of photosynthesis. Examples of these mutualistic interactions are root nodules and mycorrhizae.

3. Plants distribute water and dissolved mineral ions through water-conducting tubes of xylem, a vascular tissue. Cells called tracheids and vessel members form the tubes but are dead at maturity. Their waterproofed walls interconnect as narrow pipelines. *30.3*

4. Plants lose water by transpiration, or evaporation of water from leaves and other parts exposed to air. *30.3*

5. Here are the points of the cohesion–tension theory of water transport in plants: *30.3*

    a. Inside xylem's water-conducting cells, continuous negative pressures (tensions) extend from leaves to the roots. Transpiration causes the tension.

    b. When water molecules escape from the leaves, replacements are pulled into the leaf under tension.

    c. The collective strength of numerous hydrogen bonds between water molecules imparts cohesion that allows water molecules to be pulled up as continuous fluid columns. The "cohesion" part of the theory refers to the capacity of the hydrogen bonds between water molecules to resist rupturing when put under tension.

6. Most plants have a waxy, water-impermeable cuticle covering their aboveground parts. They lose water and take up carbon dioxide at stomata. A pair of guard cells defines each of these microscopically small openings across the epidermis of leaves and stems. Guard cells are specialized parenchyma cells. *30.4*

7. Stomata open and close at different times. Control of their action balances water conservation with carbon dioxide uptake and the release of oxygen. *30.4*

    a. In many kinds of plants, stomata open during the day. Such plants lose water but take in carbon dioxide for photosynthesis. Stomata close at night; thus, water and the carbon dioxide released by aerobically respiring cells are conserved for the next day.

    b. CAM plants open stomata and fix carbon dioxide by the C4 pathway at night. During the day, they close their stomata and use the carbon fixed the night before for photosynthesis.

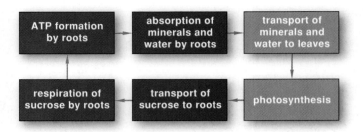

Figure 30.17 Summary of the interdependent processes that sustain the growth of vascular plants. All living cells in plants require oxygen, carbon, hydrogen, and at least thirteen mineral ions. They all produce ATP, which drives metabolic activities.

8. Plants distribute organic compounds through sieve tubes of phloem, a vascular tissue. Translocation is the distribution of organic compounds in phloem. *30.5*

9. According to the pressure flow theory, translocation is driven by differences in solute concentrations and pressure between source and sink regions. A source is any site where organic compounds are being loaded into sieve tubes (e.g., mature leaves). A sink is any site where compounds are being unloaded from sieve tubes. Roots are examples. Solute concentration gradients and pressure gradients exist for as long as companion cells expend energy to load solutes into the sieve tubes and for as long as solutes are removed at the sink. *30.5*

Review Questions

1. Define soil, then distinguish between: *30.1*
    a. humus and loam
    b. leaching and erosion
    c. macronutrient and micronutrient (for plants)

2. Define nutrient. What are signs that a plant is deficient in one of the essential nutrients listed in Table 30.1? *30.1*

3. What is the function of the Casparian strip in roots? *30.2*

4. Using Dixon's model, explain how water moves from soil upward through tall plants. *30.3*

5. Describe the structure and function of a plant cuticle. *30.4*

6. Which type of ion influences stomatal action? *30.4*

7. Explain translocation according to the pressure flow theory described in this chapter. *30.5*

Self-Quiz ANSWERS IN APPENDIX III

1. Carbon, hydrogen, oxygen, nitrogen, and potassium are examples of _____ for plants.
    a. macronutrients   d. essential elements
    b. micronutrients   e. both a and d
    c. trace elements

2. A _____ strip in endodermal cell walls forces water and solutes to move through root cells, not around them.
    a. cutin   b. lignin   c. Casparian   d. cellulose

3. The nutrition of some plants depends on a root–fungus association known as a _____ .
    a. root nodule   c. root hair
    b. mycorrhiza   d. root hypha

4. The nutrition of some plants depends on a root–bacterium association known as a _____ .
   a. root nodule
   b. mycorrhiza
   c. root hair
   d. root hypha

5. Water can be pulled up through a plant by the cumulative strength of _____ between water molecules.

6. Water evaporation from plant parts is called _____ .
   a. translocation
   b. expiration
   c. transpiration
   d. tension

7. Water transport from roots to leaves is explained by _____ .
   a. the pressure flow theory
   b. differences in source and sink solute concentrations
   c. the pumping force of xylem vessels
   d. the cohesion–tension theory

8. In daytime, most plants lose _____ and take up _____ .
   a. water; carbon dioxide
   b. water; oxygen
   c. oxygen; water
   d. carbon dioxide; water

9. At night, most plants conserve _____ , and _____ accumulates.
   a. carbon dioxide; oxygen
   b. water; oxygen
   c. oxygen; water
   d. water; carbon dioxide

10. In phloem, organic compounds flow through _____ .
    a. collenchyma cells
    b. sieve tubes
    c. vessels
    d. tracheids

11. Match the concepts of plant nutrition and transport.
    _____ stomata
    _____ nutrient
    _____ sink
    _____ root system
    _____ hydrogen bonds
    _____ transpiration
    _____ translocation

    a. evaporation from plant parts
    b. response to scarce soil nutrients
    c. balancing water loss with carbon dioxide requirements
    d. cohesion in water transport
    e. sugars unloaded from sieve tubes
    f. organic compounds distributed through the plant body
    g. element with roles in metabolism that no other element can fulfill for an organism

Critical Thinking

1. Home gardeners, like farmers, must ensure that their plants have access to nitrogen from either nitrogen-fixing bacteria or fertilizer. Insufficient nitrogen stunts plant growth; leaves turn yellow and die. Which major classes of biological molecules incorporate nitrogen? How would a low nitrogen level in plants affect biosynthesis and cause symptoms of nitrogen deficiency?

2. When moving a plant from one place to another, it helps to include some native soil around the roots. Explain why, given what you know about mycorrhizae and root hairs.

3. In the sketch at *right*, label the actual stoma. Now think about how Henry discovered a way to keep all of a plant's stomata open at all times. He also figured out how to keep those of another plant closed all the time. Both plants died. Explain why.

4. Allen is studying the rate of transpiration from tomato plant leaves. He notices that several environmental factors, including wind and relative humidity, affect the rate. Explain why.

5. You have just returned home from a three-day vacation. Your plants tell you, by their severe wilting, that you forgot to water them before you departed. Being aware of the cohesion–tension theory of water transport, explain what happened to them.

6. Not having a green thumb, your friend Stephanie decides to grow zucchini plants, which are very forgiving of poor soils

Figure 30.18 Middle fork of the Salmon River, Idaho.

and amateur gardeners. She plants too many seeds and ends up with far too many plants. Among them you happen to notice a stunted plant and decide to find out what happened to it. After many experiments, you decide that its leaves are producing a mutated, malfunctioning form of an enzyme that is necessary for the formation of sucrose. Knowing what you do about the pressure flow theory, explain why plant growth was hindered.

7. At the middle fork of the Salmon River in Idaho, clear water from a wilderness area (visible at *right* in Figure 30.18) converges on brown-colored water (visible at *left* and *center*). The brown water is enriched with silt from a wilderness area that was disturbed by cattle ranching and some other human activities. Knowing what you do about the nature of topsoil, what is probably happening to plant growth in the disturbed habitat? Knowing how fishes acquire oxygen for aerobic respiration, how might the silt be affecting their survival?

Selected Key Terms

CAM plant *30.4*
carnivorous plant *CI*
Casparian strip *30.2*
cohesion–tension theory *30.3*
companion cell *30.5*
cuticle *30.4*
cutin *30.4*
endodermis *30.2*
erosion *30.1*
exodermis *30.2*
guard cell *30.4*
humus *30.1*
leaching *30.1*
loam *30.1*
mutualism *30.2*
mycorrhiza *30.2*
nitrogen fixation *30.2*
nutrient *30.1*
phloem *30.5*
plant physiology *CI*
pressure flow theory *30.5*
root hair *30.2*
root nodule *30.2*
sieve tube *30.5*
soil *30.1*
stoma (stomata) *30.4*
topsoil *30.1*
tracheid *30.3*
translocation *30.5*
transpiration *30.3*
turgor pressure *30.4*
vascular cylinder *30.2*
vessel member *30.3*
xylem *30.3*

Readings

Hausenbuiller, R. 1985. *Soil Science: Principles and Practices.* Third edition. Dubuque, Iowa: W. C. Brown.

Hopkins. W. G. 1995. *Introduction to Plant Physiology.* New York: Wiley.

Raven, R., R. Evert, and S. Eichhorn. 1999. *Biology of Plants.* Sixth edition. New York: Freeman.

Salisbury, F., and C. Ross. 1992. *Plant Physiology.* Fourth edition. Belmont, California: Wadsworth.

PLANT REPRODUCTION

A Coevolutionary Tale

We find flowering plants almost everywhere, from icy tundra to deserts to oceanic islands. What accounts for their distribution and diversity? Consider the **flower**, a specialized reproductive shoot. About 435 million years ago, when plants first invaded the land, insects that ate decaying plant parts and spores probably weren't far behind. The plant smorgasbord seems to have favored natural selection of winged insects with an amazing array of sucking, piercing, and chewing mouthparts.

By 390 million years ago, in humid coastal forests, seed-bearing plants were making pollen grains. These tiny, sperm-bearing packages can travel to eggs, which develop in ovules in female plant parts. Pollen is rich in nutrients. At first, air currents may have dispersed it to the ovules. Then hungry insects made the connection between "plant parts with pollen" and "food." Plants lost some pollen to insects but gained a reproductive advantage. How? Unlike air currents, pollen-dusted insects clambering over plants could deliver pollen right to the ovules.

Plant structures evolved in novel or modified ways that were more enticing to insects, so pollen dispersal became more efficient. Insects that became specialized in detecting and gathering pollen from a specific source gained a competitive edge over insects that spent more time and energy on random searches for food.

What we are describing is a case of **coevolution**. The word refers to two or more species jointly evolving as an outcome of close ecological interactions. A heritable change in one species affects selection pressure operating between them, so the other species evolves, too.

In this case, the more enticing plants enjoyed more home deliveries, produced more seeds, and *improved their chances of reproductive success*. And so, over time, plants with distinctive flowers, fragrances, and sugar-rich nectar coevolved with pollinators.

A **pollinator** is any agent that transfers pollen from male to female reproductive parts of flowers of the same plant species. Besides insects, pollinating agents include air currents, water currents, bats, birds, and other animals (Figures 31.1 and 31.2).

You can correlate many floral features with specific mutualists. For instance, a flower's reproductive parts are positioned so that pollinating agents will brush past them. Many parts are positioned above nectar-filled floral tubes the same length as the feeding device of a preferred pollinator. Red and yellow flowers attract birds, which have great daytime vision but a poor sense of smell. As you might suspect, plants that birds visit don't divert metabolic resources to making fragrances. Red flowers don't attract beetles. Neither do flowers with nectar cups large and deep enough for beetles to

Figure 31.1 Example of adaptations uniting a flowering plant with its pollinator. The giant saguaro of Arizona's Sonoran Desert has large, showy white flowers at the tips of its spiny arms. Insects and birds visit the flowers by day, and bats visit by night. The plant offers these animals nectar. And the animals transport pollen grains, which stick to their body, from one cactus plant to another.

How we see it   How bees see it

Figure 31.2 (**a**) Bahama woodstar sipping nectar from a hibiscus blossom. Like other hummingbirds, it forages for nectar in midflight. Its long, narrow bill coevolved with long, narrow floral tubes. (**b**,**c**) Shine ultraviolet light on a gold-petaled marsh marigold to reveal its bee-attracting pattern.

drown in. Like flies, beetles pollinate flowers that smell like rotten meat, moist dung, or decaying litter on the forest floor—where beetles first evolved.

Daisies and other fragrant flowers with distinctive patterns, shapes, and red or orange components attract butterflies, which forage by day. Nectar-sipping bats and most moths forage by night. They pollinate intensely sweet-smelling flowers with white or pale petals that are more visible than colored petals in the dark. Long, thin mouthparts of moths and butterflies reach nectar in narrow floral tubes or floral spurs. The Madagascar hawkmoth uncoils a mouthpart the same length as a narrow floral spur of an orchid, *Angraecum sesquipedale*. It is 22 centimeters (more than 8–1/2 inches) long!

Flowers with sweet odors and yellow, blue, or purple parts attract bees. Their ultraviolet-light-absorbing pigments form patterns that say, "Nectar here!" Unlike us, bees see ultraviolet light and find nectar guides alluring (Figure 31.2b,c). Unlike beetles, they also have long mouthparts that extend into floral tubes.

In short, flowers contribute to reproductive success of the plants that bear them. This chapter focuses on the reproductive modes and development of these plants.

Key Concepts

1. Sexual reproduction is the premier reproductive mode of flowering plant life cycles. It involves the formation of spores and gametes, both of which develop inside the specialized reproductive shoots called flowers.

2. Microspores form in a flower's male reproductive parts and develop into pollen grains, the sperm-bearing male gametophytes. Animals, air currents, and other pollinating agents transfer pollen grains to the female reproductive parts of flowers.

3. Megaspores develop in ovules, structures that form on the inner ovary wall of the flower's female reproductive parts. Within each ovule, a mature female gametophyte forms from a megaspore. One of its cells is the egg.

4. After sperm fertilize the eggs, ovules mature into seeds. Each seed consists of an embryo sporophyte and various tissues that function in its nutrition and protection.

5. While seeds are developing, tissues of the ovary and sometimes neighboring tissues mature into fruits, which function in seed dispersal. Air currents, water currents, and animals function as dispersal agents.

6. Many species of flowering plants also reproduce asexually by mechanisms of vegetative growth, parthenogenesis, and tissue culture propagation.

REPRODUCTIVE STRUCTURES OF FLOWERING PLANTS

Think Sporophyte and Gametophyte

Like us, flowering plants engage in sex. They produce, nourish, and protect sperm and eggs in reproductive systems. Like human females, they nourish and protect developing embryos. They put out floral invitations to third parties—pollinators that get sperm to eggs. Long before we thought of it, they were using perfumes and colors to improve the odds for sexual success.

When you hear the word "plant," you may think of something like a cherry tree (Figure 31.3*a*). The tree is a typical **sporophyte**, a vegetative body that grows, by mitotic cell divisions, from a fertilized egg. During the life cycle, this sporophyte bears flowers—floral shoots specialized for reproduction. In flowers, haploid spores give rise to haploid bodies called **gametophytes** (Section 10.5). Sperm develop inside male gametophytes, and the eggs develop inside female gametophytes. Fusion of a haploid sperm with a haploid egg at fertilization results in a cell with two sets of genetic instructions.

Components of Flowers

Flowers form at floral shoots of the primary plant body. As they form, they differentiate into sepals and petals (nonfertile parts) and stamens and carpels (fertile parts). Figures 31.3*b* and 31.4 show how the parts are arranged. All connect to a receptacle, the modified base of a floral shoot. Peel open a rosebud. Sepals, the outer whorl of leaflike parts, surround leaflike petals, which surround the fertile parts. This is a common pattern. Collectively, sepals are a flower's calyx, and petals are its corolla.

Like leaves, sepals and petals have ground tissues, vascular tissues, and epidermis. Why are many flowers sweet smelling? Some epidermal cells of petals produce fragrant oils. What gives the petals their shimmer and color? Cells of the ground tissue contain pigments, such as carotenoids (yellow to red-orange) and anthocyanins (red to blue), plus small, light-refracting crystals. What does a flower's fragrance or its coloration, patterning, or arrangement of petals do? It attracts pollinators.

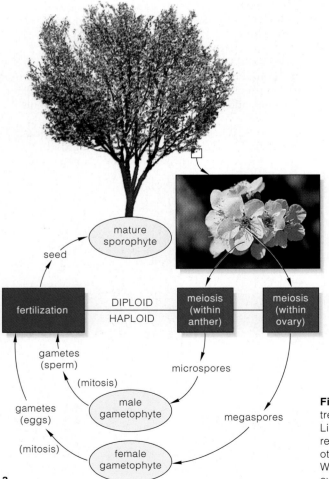

a

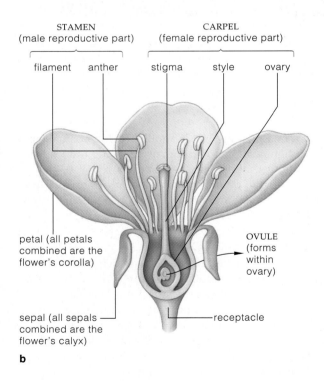

b

Figure 31.3 (**a**) Overview of flowering plant life cycles, using a cherry tree (*Prunus*) as the example. (**b**) Structure of one flower, a cherry blossom. Like the flowers of many plants, this blossom has a single carpel (a female reproductive part) and stamens (the male reproductive parts). Flowers of other plants have two or more carpels, often fused into a single structure. Whether single or fused, carpels have an ovary and a stigma. Often the ovary extends upward as a slender column, called the style.

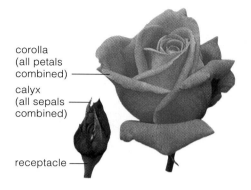

Figure 31.4 Location of floral parts in a prized cultivated plant, the rose (*Rosa*).

corolla (all petals combined)

calyx (all sepals combined)

receptacle

Where Pollen and Eggs Develop

Nearly all **stamens**, a flower's male reproductive parts, consist of an anther and a one-veined stalk (filament). An anther is divided into pollen sacs, the chambers in which walled, haploid spores form and give rise to the male gametophytes called **pollen grains** (Figure 31.5).

A flower's female reproductive parts are located at its center. You may have heard someone call these parts pistils, but their more recent name is **carpels**.

Many flowers have one carpel. Others have more, often fused as a compound structure. The lower part of single or fused carpels—the **ovary**—is where the eggs develop, fertilization takes place, and seeds mature. The more formal name for flowering plants, angiosperm, refers to the carpel (from the Greek *angeion*, meaning vessel, and *sperma*, seed). The upper portion of a carpel is the stigma. This sticky or hairy surface tissue captures pollen grains and favors their germination. Commonly the stigma is elevated upon a style, a slender extension of the upper ovary wall (Figure 31.3*b*).

Not every flower has stamens and carpels. Some species have *perfect* flowers, with both female and male parts. Others have *imperfect* flowers; they have female or male parts. In certain species such as oaks, the same individual has male and female flowers. In willows and other species, they are on separate, individual plants.

Sexual reproduction is the dominant reproductive mode of flowering plant life cycles. Such cycles alternate between the production of sporophytes (spore-producing bodies) and gametophytes (gamete-producing bodies).

Flowering plant sporophytes form from a fertilized egg by mitotic cell divisions and cell growth. A sporophyte is all of the plant body except for its male and female gametophytes.

Gametophytes arise from haploid spores, which form in stamens and carpels—the male and female reproductive parts of the flower.

Each male gametophyte gives rise to a sperm-producing pollen grain. Each female gametophyte develops inside a carpel's ovary; one of its cells is an egg. Also inside the ovary, fertilization takes place and seeds mature.

Pollen Sets Me Sneezing

In the year 1835 Sidney Smith wrote, "I am suffering from my old complaint, the hay-fever . . . that sets me sneezing; and if I begin sneezing at twelve, I don't leave off till two o'clock, and am heard distinctly in Taunton [six miles distant] when the wind sets that way."

Sidney suffered what is now called *allergic rhinitis*. This is the name for hypersensitivity to a normally harmless substance. Such hypersensitivity is common.

The pollen of ragweed, shown in Figure 31.5, brought on Sidney's dreaded hay fever, but it wasn't out to get him. Every spring and summer, flowering plants release pollen grains. In many millions of people, white blood cells respond by mounting an immune response against some proteins that project from the surface of pollen grain walls. The pollen grains from different plants have differences in their wall proteins. The immune system of a person who is hypersensitive might chemically respond to some of these but not others. The misdirected response results in a profusely runny nose, reddened and itchy eyelids, congestion, and bouts of sneezing.

Hay fever is a genetic abnormality; it runs in families. Some individuals simply are genetically predisposed to overreact to some kinds of pollen. Infections, emotional stress, and changes in temperature may also trigger overreactions to pollen.

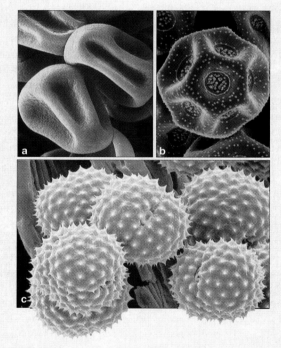

Figure 31.5 Pollen grains of (**a**) grass, (**b**) chickweed, and (**c**) ragweed plants. Pollen grains of most families of plants differ in size, wall sculpturing, and number of wall pores.

31.3

A NEW GENERATION BEGINS

From Microspores to Pollen Grains

We turn now to pollen grain formation. While anthers are growing, four masses of spore-producing cells form by mitotic cell divisions. Walls develop around them. Each anther now has four chambers, called pollen sacs (Figure 31.6a). Haploid **microspores** form when cells in the sacs undergo meiosis and cytoplasmic division. They go on to develop an elaborately sculpted wall. The walled microspores divide once or twice, by mitosis, and then form pollen grains. These enter a period of arrested growth and in time will be released from the anther. Components of the pollen grain wall will help protect them from decomposers in their surroundings.

As soon as many types of pollen grains form, they produce sperm nuclei, which are the male gametes of flowering plants. Other types don't do this until after they reach a carpel and start growing toward its ovule. In short, each pollen grain is a mature *or* an immature male gametophyte, depending on the plant species.

From Megaspores to Eggs

Meanwhile, one or more masses of cells are forming on the inner wall of the flower's ovary. Each is the start of an **ovule**, a structure that houses a female gametophyte and that may become a seed. As each cell mass grows, a tissue forms within it, and one or two protective layers called **integuments** form around it. Inside the mass, a cell divides by meiosis, and four haploid spores form. Spores formed in flowering plant ovaries are generally larger than microspores and are called **megaspores**.

Commonly, all megaspores but one disintegrate. That one undergoes mitosis three times without cytoplasmic division. At first, it has eight nuclei (Figure 31.6b). Each nucleus migrates to a prescribed location, and then the cytoplasm divides. The result, a seven-celled embryo sac, is a female gametophyte. One cell has two nuclei. This endosperm mother cell will help form the **endosperm**, a nutritive tissue for the embryo. Another cell is the egg.

From Pollination to Fertilization

Flowering plants release pollen in spring. You are very aware of this reproductive event if you are one of the millions of people who experience hay fever, an allergic reaction to wall proteins of pollen grains (Section 31.2).

Specifically, **pollination** is a transfer of pollen grains to a receptive stigma. Air or water currents, birds, bats, insects, and many other agents of pollination carry out such transfers, as you read in the chapter introduction.

Once a pollen grain lands on a receptive stigma, it germinates. For this event, germination means a pollen grain resumes growth and then develops into a tubular

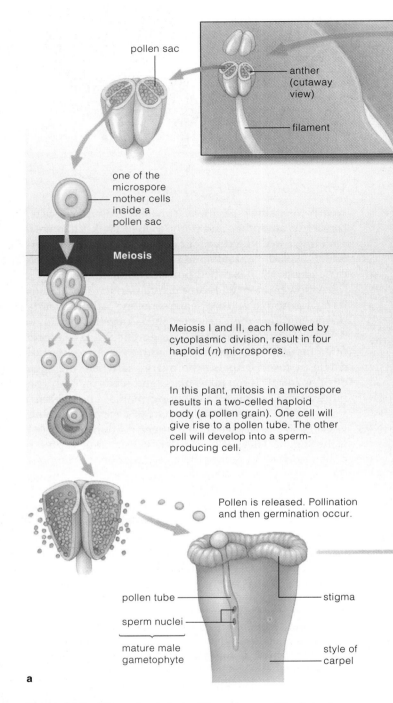

Figure 31.6 Life cycle of cherry (*Prunus*), one of the flowering plants classified as a eudicot (true dicot). (**a**) How pollen grains develop and germinate. (**b**) Events in this ovary's ovule.

structure. This "pollen tube" grows down through the tissues of the ovary and carries the sperm nuclei with it (Figure 31.6b). Chemical and molecular cues guide the growth of the pollen tube through the ovary's tissues,

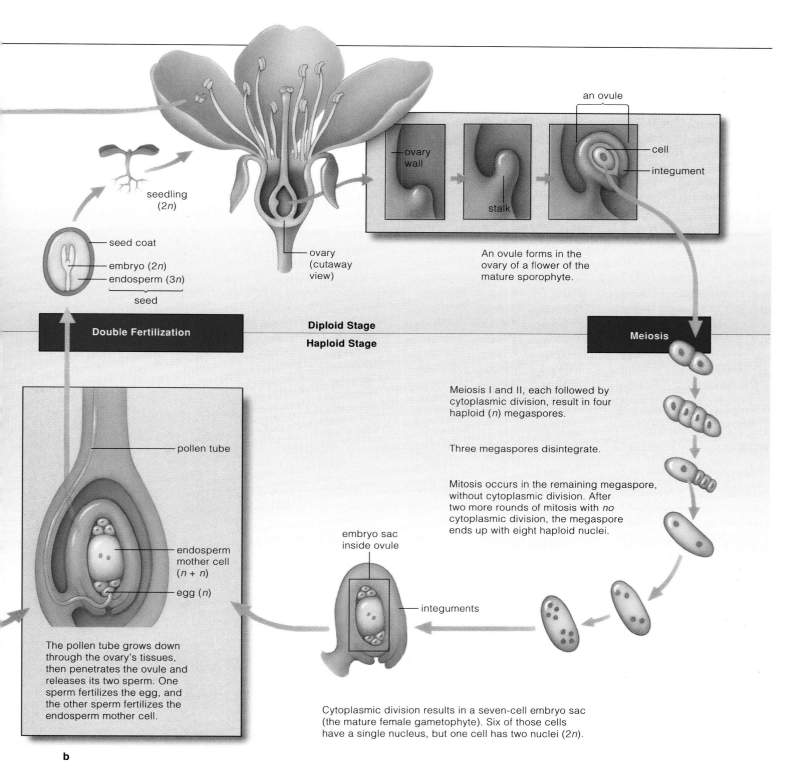

seedling
(2n)

seed coat

embryo (2n)

endosperm (3n)

seed

ovary
(cutaway
view)

an ovule

ovary
wall

stalk

cell

integument

An ovule forms in the
ovary of a flower of the
mature sporophyte.

Double Fertilization

Diploid Stage

Haploid Stage

Meiosis

Meiosis I and II, each followed by
cytoplasmic division, result in four
haploid (n) megaspores.

Three megaspores disintegrate.

Mitosis occurs in the remaining megaspore,
without cytoplasmic division. After
two more rounds of mitosis with *no*
cytoplasmic division, the megaspore
ends up with eight haploid nuclei.

pollen tube

endosperm
mother cell
(n + n)

egg (n)

embryo sac
inside ovule

integuments

The pollen tube grows down
through the ovary's tissues,
then penetrates the ovule and
releases its two sperm. One
sperm fertilizes the egg, and
the other sperm fertilizes the
endosperm mother cell.

Cytoplasmic division results in a seven-cell embryo sac
(the mature female gametophyte). Six of those cells
have a single nucleus, but one cell has two nuclei (2n).

b

toward the egg chamber and sexual destiny. When the
pollen tube reaches an ovule, it penetrates the embryo
sac, and its tip ruptures to release two sperm.

"Fertilization" generally means fusion of a sperm
nucleus with an egg nucleus. But **double fertilization**
occurs in flowering plants. In species having a diploid
chromosome number, one sperm nucleus fuses with an
egg nucleus to form a diploid (2n) zygote. Meanwhile,
the other sperm nucleus fuses with both nuclei of the

endosperm mother cell. The resulting cell has a triploid
(3n) nucleus and will give rise to the nutritive tissue.

**In flowering plants, sperm cells form within pollen grains,
the male gametophytes. Ovules form inside ovaries, and
female gametophytes (with egg cells) develop within them.**

**After pollination and double fertilization, an embryo and
nutritive tissue form in the ovule, which becomes a seed.**

FROM ZYGOTES TO SEEDS AND FRUITS

After fertilization, the newly formed zygote embarks on a course of mitotic cell divisions that lead to a mature embryo sporophyte. It develops as part of an ovule and is accompanied by formation of a **fruit**: a mature ovary, with or without other, neighboring tissues.

Formation of the Embryo Sporophyte

Let's look at how a shepherd's purse (*Capsella*) embryo forms. By the time it reaches the stage in Figure 31.7*e*, two **cotyledons**, or seed leaves, have started to develop from lobes of meristematic tissue. Cotyledons are part of all flowering plant embryos. Dicot embryos have two, and monocot embryos have one. The *Capsella* embryo, like those of many dicots, absorbs nutrients from the endosperm and stores them in its cotyledons. By contrast, in corn, wheat, and most other monocots, endosperm is not tapped until a seed germinates. Digestive enzymes get stockpiled in a monocot embryo's thin cotyledons. When activated, the enzymes will help transfer stored endosperm to the growing seedling.

Table 31.1 *Categories and Examples of Fruits*

SIMPLE FRUITS From one ovary of one flower.

Dry fruit

> *Dehiscent.* Fruit wall splits along definite seams to release seeds. Legume (e.g., pea, bean), poppy, larkspur, mustard
>
> *Indehiscent.* Fruit wall does not split on seams to release seeds. Acorn, grains (e.g., corn), sunflower, carrot, maple

Fleshy fruit

> *Berry.* Compound or simple ovary, many seeds. Tomato, grape, banana
>
> > Pepo. Ovary wall has hard rind. Cucumber, watermelon
> > Hesperidium. Ovary wall has leathery rind. Orange, lemon
>
> *Drupe.* One or two seeds. Thin skin, part of flesh around a seed enclosed in hardened ovarian tissue. Peach, cherry, apricot, almond, olive

AGGREGATE FRUITS

Many ovaries of one flower, all attached to the same receptacle. Many seeds. Raspberry, blackberry (not really "berries")

MULTIPLE FRUITS

Combined from ovaries of many flowers. Pineapple, fig

ACCESSORY FRUITS

Most tissues of the flesh are not derived from the ovary; e.g., mainly from the receptacle. Pome (apple, pear), strawberry

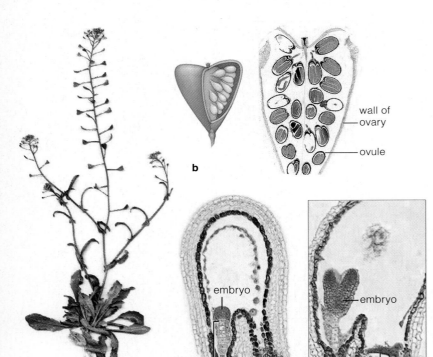

wall of ovary

ovule

b

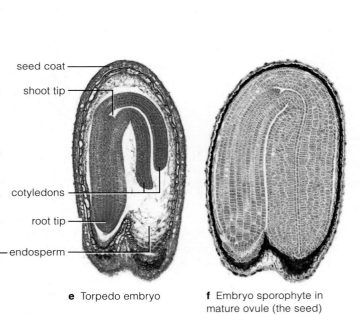

seed coat

shoot tip

cotyledons

root tip

embryo

embryo

endosperm

suspensor

a

c Globular embryo   **d** Heart-shaped embryo   **e** Torpedo embryo   **f** Embryo sporophyte in mature ovule (the seed)

Figure 31.7 Dicot seed development. (**a**) Mature sporophyte of shepherd's purse (*Capsella*). (**b**) Seeds in ovary. (**c,d**) The suspensor transfers nutrients to the early embryo sporophyte from the parent plant. The embryo is well developed in (**e**). It is mature in (**f**). Micrographs are not to the same scale.

fruit wall | cotyledons

Figure 31.8 From flowers to fruits—three examples. (**a**–**d**) Strawberry (*Fragaria*), a fragrant accessory fruit. Most of the fleshy tissues arise from the flower's receptacle. Numerous fruits are perched on the surface of the mature receptacle; each has a hard fruit wall around the embryo sporophyte.

(**e**) Pineapple (*Ananas*), a multiple fruit. Native Americans cultivated pineapples. To them, it was a symbol of hospitality. They used the juice as a base for an alcoholic beverage. Christopher Columbus thought the fruit looked like pine cones, hence the name.

(**f**–**i**) Fruit formation on an apple (*Malus*) tree. When the petals drop, this is a sign that the eggs have become fertilized.

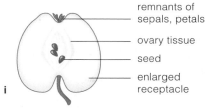

remnants of
sepals, petals

ovary tissue

seed

enlarged
receptacle

i

Seeds and Fruit Formation

From the time a zygote forms inside an ovule until it has become a mature embryo sporophyte, the parent plant transfers nutrients to ovule tissues. Food reserves accumulate in endosperm or in cotyledons. The ovule eventually separates from the wall of the ovary, and its integuments thicken and harden into a seed coat. The embryo, food reserves, and coat are a self-contained package—a **seed**, which we define as a mature ovule.

When seeds form, other floral parts change and start to form fruits. As Table 31.1 shows, three categories are *simple* fruits (either dry or fleshy, and derived from a single ovary), *aggregate* fruits (from numerous separate ovaries of one flower), and *multiple* fruits (from many separate ovaries of a single flower, all attached to the same receptacle).

At maturity, acorns and some other simple dry fruits are intact; pea pods unzip to release seeds. Peaches, as in Section 23.5, are simple fleshy fruits. Strawberries are an aggregate fruit that has many seeds on a fleshy, expanded receptacle (Figure 31.8*a–d*). Pineapples are a cone-shaped multiple fruit. They form when ovaries of many separate flowers fuse together into a fleshy mass as they enlarge. A waxy, hardened rind covers the fruit (Figure 31.8*e*). Apples, an accessory fruit, are mainly an enlarged receptacle and calyx (Figure 31.8*f–i*).

Fleshy fruits have three tissue divisions, although sometimes the three are not immediately visible. The *endo*carp is the innermost portion around the seed or seeds. *Meso*carp is the fleshy portion. *Exo*carp is the skin. Together, the three fruit regions are a **pericarp**. In acorns and other dry fruits, the pericarp is typically thin. In true berries, such as tomatoes and grapes, the pericarp is thin skinned and relatively soft at maturity.

A mature ovule, which encases an embryo sporophyte and food reserves inside a protective coat, is a seed.

A mature ovary, with or without additional floral parts that have become incorporated into it, is a fruit.

Fruits are classified by whether they are dry or fleshy, are derived from one or more ovaries, and incorporate other tissues besides those of the ovary.

Fleshy fruits have three regions. Innermost is the endocarp, which surrounds the seeds. The mesocarp is the fleshy part, and the exocarp is the skin. Together, the three regions are known as the pericarp.

DISPERSAL OF FRUITS AND SEEDS

The fruits you read about in the preceding section have a common function: *seed dispersal*. To gain insight into their structure, think about how they coevolved with particular dispersing agents—currents of air or water, or animals passing by.

Consider the wind-dispersed fruit of maples (*Acer*), as shown in Figure 31.9*a*. The pericarp of a maple fruit extends out like wings. When the fruit breaks in half and drops from a tree, it interacts with air currents and spins sideways. The wings whirl the fruit far enough away that the embryo sporophytes inside seeds will not have to compete with the parent plant for water, minerals, and sunlight.

Brisk winds have been known to transport fruits as far as ten kilometers from a tree. Air currents easily lift the tiny fruit of dandelions by their outward-pluming "parachute." And air currents readily disperse orchid seeds, which are as fine as dust particles.

Other fruits taxi to new locations on or in animals, including many birds, mammals, and insects. Some can adhere to feathers, feet, or fur with hooks, spines, hairs, and sticky surfaces. Fruits of cocklebur, bur clover, and bedstraw are like this. Embryo sporophytes tucked in the seed coats of fleshy fruits survive being eaten and assaulted by the digestive enzymes in an animal's gut. Besides digesting the fruit's flesh, these enzymes digest some of the seed coat. The digested portions make it easier for the embryo to break through a hard coat after the seeds are expelled from the animal body in feces and start to germinate. Germination is a topic that will be described in the next chapter.

Some water-dispersed fruits have heavy wax coats, and others, including sedges, have sacs of air that help them float. With its thick, water-repelling pericarp, the coconut palm fruit is adapted to travel across the open ocean. If it doesn't wash onto a beach soon enough, saltwater might penetrate the pericarp and kill the embryo. Coconuts can drift hundreds of kilometers before this happens.

a   wing   seed (in carpel)   b

Figure 31.9 (**a**) Winged seeds of maple (*Acer*). (**b**) A few students taking a chocolate ice cream break and indirectly contributing to the long-term reproductive success of cacao (*Theobroma cacao*). (**c**) Cacao fruit, or pod. Each contains up to forty seeds ("beans"). Seeds are processed into cocoa butter and essences of chocolate products. Interactions among at least a thousand compounds in chocolate exert compelling (some say addictive) effects on the human brain. The average American buys 8 to 10 pounds of chocolate annually. The average Swiss citizen craves a whopping 22 pounds. With that kind of demand, *T. cacao* seeds are cherished and "dispersed" in orderly arrays in plantations, where the trees that grow from them are carefully tended.

Humans are grand agents of dispersal. In the past, explorers transported seeds all over the world, although most places now control imports. Cacao (Figure 31.9*c*), oranges, corn, and other plants were domesticated and encouraged to reproduce in new habitats, in far greater numbers than they did on their own. And in evolutionary terms, reproductive success is what life is all about.

Seeds and fruits are structurally adapted for dispersal by air currents, water currents, and many kinds of animals.

Why So Many Flowers and So Few Fruits?

In the Sonoran Desert of Arizona (Figure 31.10*a*), the giant saguaro produces as many as a hundred flowers at the tips of its huge, spiny arms. Remember the large, showy flowers in Figure 31.1? Each flower blooms for only twenty-four hours. Again, insects visit the flowers by day, and bats by night. The animals benefit by getting nectar, and the giant saguaro benefits because the animals deliver their dusting of pollen grains to other giant saguaros.

The flowers that do get pollinated, courtesy of insects or bats, now begin the task of producing seeds and fruits. Petals wilt, then they wither as the many egg cells inside the ovary become fertilized. Each egg-containing ovule expands and matures into a seed. During the same interval, ovaries develop into fruits (Figure 31.10). Weeks later, the plum-sized fruits split open, exposing a bright red interior and dark seeds.

White-winged doves feast on the ripe fruits. When they fly away, each might carry hundreds of seeds in their gut. A few seeds may elude the gut's digestive enzymes and later end up on the ground. There, each seed will have a tiny chance of growing into a giant saguaro.

commonly produce far more flowers than they do mature fruits. They seem to be producing flowers excessively and giving up chances to make seeds and leave descendants. Doing so is not a reproductive adaptation that you would expect, based on Darwinian evolutionary theory.

Well, *suppose that some of the plants simply did not receive enough pollen to have all their egg cells fertilized.* After all, unfertilized flowers cannot produce seeds and fruits. This hypothesis has been experimentally tested for some plant species. Researchers carefully brushed pollen on every single flower. In many instances, the plants still failed to produce fruit for every flower!

Alternatively, *suppose the presumed "excess" flowers are formed strictly to produce pollen for export to other plants.*

Figure 31.10 (**a**) Giant saguaro growing in Arizona's Sonoran Desert. (**b**) Two fruits. The one above set, and the one below it failed to set. (**c**) A red, maturing fruit.

One of the puzzling aspects of this annual event is the frequency with which saguaro flowers fail to give rise to fruit (Figure 31.10*b*, for example). Why would a cactus invest so much energy constructing a hundred flowers if only thirty or so of them will set fruit? Compare one of these "inefficient" plants with a species that produces a hundred flowers, all of which develop into fruit. Which do you think will leave more descendants?

Common sense might lead you to conclude, "The more fruits, the better." Yet saguaros—like many other plants—

Pollen grains are small. Energetically speaking, they are inexpensive to produce compared to large, calorie-rich, seed-containing fruits. Thus, for a fairly small investment, a plant might reap a large reward in offspring that carry its genes.

The idea that some kinds of plants actually set aside flowers exclusively for pollen export is only now being tested for saguaros. Perhaps you might like to design and carry out experiments that will help clear up the mystery of the "excess" flowers of saguaros and similar plants.

ASEXUAL REPRODUCTION OF FLOWERING PLANTS

Asexual Reproduction in Nature

The features of sexual reproduction that we considered in earlier sections dominate flowering plant life cycles. Bear in mind, many species also reproduce asexually by modes of **vegetative growth**, as listed in Table 31.2. In essence, new roots and shoots grow right out from extensions or fragments of parent plants. Such asexual reproduction proceeds by way of mitosis. This means a new generation of offspring is genetically identical to the parent; it is a clone.

One "forest" of quaking aspen (*Populus tremuloides*) provides us with an impressive example of vegetative reproduction. The leaves of this flowering plant species tremble even in the slightest breeze, hence the name. Figure 31.11 is a panoramic view of the shoot systems of one individual. The root system of the parent plant gave rise to adventitious shoots, which became separate shoot systems. Barring rare mutations, individuals at the north end of this vast clone are genetically identical to individuals at the south end. Water travels from roots near a lake all the way to shoot systems in much drier soil. Dissolved ions travel in the opposite direction.

As long as environmental conditions favor growth and regeneration, such clones are about as close as one can get to being immortal. No one knows how old the aspen clones are. The oldest known clone is a ring of creosote bushes (*Larrea divaricata*) growing in the Mojave Desert. It has been around for the past 11,700 years.

The reproductive possibilities are amazing. Watch strawberry plants send out horizontal, aboveground stems (runners), then watch the new roots and shoots develop at every other node. The oranges you eat may be descended from a single tree in southern California that reproduced by **parthenogenesis**. With this type of process, an embryo develops from an unfertilized egg.

Parthenogenesis can be stimulated when pollen has contacted a stigma even though a pollen tube has not grown through the style. Maybe certain hormones from the stigma or pollen grains diffuse to the unfertilized egg and trigger formation of an embryo. That embryo is $2n$ as a result of fusion of the products of mitotic cell division in the egg. A $2n$ cell outside the gametophyte also may be stimulated to develop into an embryo.

Induced Propagation

Most houseplants, woody ornamentals, and orchard trees are clones. People typically propagate them from cuttings or fragments of shoot systems. For example, with suitable encouragement, a severed African violet leaf may form a callus from which adventitious roots develop. A callus is one type of meristem. Remember, a meristem is a localized region of plant cells that retain the potential for mitotic cell division.

Or consider a twig or bud from one plant, grafted onto a different variety of some closely related species. Vintners in France, for instance, graft prized grapevines onto disease-resistant root stock from America.

Frederick Steward and his colleagues were pioneers in **tissue culture propagation**. They cultured small bits of phloem from the differentiated roots of carrot plants (*Daucus carota*) in rotating flasks. They used a liquid growth medium that contained sucrose, minerals, and vitamins. The liquid also contained coconut milk, which Steward knew was rich in then-unidentified growth-inducing substances. As the flasks rotated, individual cells that were torn away from the tissue bits divided and formed multicelled clumps, which sometimes gave rise to new roots (Figure 31.12a). Steward's experiments were among the first to demonstrate that some cells of

Table 31.2 Asexual Reproductive Modes of Flowering Plants		
Mechanism	Examples	Characteristics
VEGETATIVE REPRODUCTION ON MODIFIED STEMS		
1. Runner	Strawberry	New plants arise at nodes along aboveground horizontal stems.
2. Rhizome	Bermuda grass	New plants arise at nodes of underground horizontal stems.
3. Corm	Gladiolus	New plants arise from axillary buds on short, thick, vertical underground stems.
4. Tuber	Potato	New shoots arise from axillary buds (tubers are the enlarged tips of slender underground rhizomes).
5. Bulb	Onion, lily	New bulbs arise from axillary buds on short underground stems.
PARTHENOGENESIS		
	Orange, rose	An embryo develops without nuclear or cellular fusion (for example, from an unfertilized haploid egg or by developing adventitiously, from tissue surrounding the embryo sac).
VEGETATIVE PROPAGATION		
	Jade plant, African violet	A new plant develops from tissue or structure (a leaf, for instance) that drops from the parent plant or is separated from it.
TISSUE CULTURE PROPAGATION		
	Orchid, lily, wheat, rice, corn, tulip	A new plant is induced to arise from a parent plant cell that has not become irreversibly differentiated.

Figure 31.11 A mere portion of Pando the Trembling Giant, so named by Michael Grant and his coworkers at the University of Colorado, who studied its genetic makeup. This "forest" in Utah is actually a single asexually reproducing male organism, of a type called quaking aspen (*Populus tremuloides*). Its root system extends through about 106 hectares (about 262 acres) of soil and functionally supports its 47,000 genetically identical shoots; that is, the trees. By one estimate, this clone weighs more than 5,915,000 metric tons.

Figure 31.12 (**a**) Cultured cells from a carrot plant. At the time this photograph was taken, roots and shoots of embryonic plants were already forming. (**b**) Young orchid plants, developed from cultured meristems and young leaf primordia. Orchids are one of the most highly prized cultivated plants. Before the meristems were cloned, they were difficult to hybridize. From the time seeds form, it can take seven years or more until a new plant bears flowers. In natural habitats, the seeds normally will not germinate unless they interact with a specific fungus.

specialized tissues still house the genetic instructions required to produce an entire individual.

Researchers now use shoot tips and other parts of individual plants for tissue culture propagations. The techniques prove useful when an advantageous mutant arises. Such a mutant may show resistance to a disease that is crippling to wild-type plants of the same species. Tissue culture propagation can result in hundreds, even thousands of identical plants from merely one mutant specimen. The techniques are already being employed in efforts to improve major food crops, including corn, wheat, rice, and soybeans. They also are being used to increase production of hybrid orchids, lilies, and other prized ornamental plants (Figure 31.12*b*).

Besides reproducing sexually, flowering plants engage in asexual reproduction, as by vegetative growth.

SUMMARY　　　　　　　　*Gold* indicates text section

1. Sexual reproduction is the main reproductive mode of flowering plant life cycles. Diploid sporophytes, or spore-producing plant bodies, develop, as do haploid gametophytes (gamete-producing bodies). *31.1*

  a. The sporophyte is a multicelled vegetative body with roots, stems, leaves, and, at some point, flowers. Flowers of many species coevolved with animals that feed on nectar or pollen and also serve as pollinators. Gametophytes form in male and female floral parts.

  b. Many flowering plants also reproduce asexually. They can do this naturally (as by runners, rhizomes, and bulbs) and artificially (as by cuttings and grafting).

2. Flowers typically have sepals, petals, and one or more stamens, which are male reproductive structures. They also have carpels, the female reproductive structures. Most or all of the reproductive structures are attached to a receptacle, the modified end of a floral shoot. *31.1*

  a. Anthers of stamens contain pollen sacs in which cells divide by meiosis. A wall develops around each resulting haploid cell (a microspore), which develops into a sperm-bearing pollen grain (male gametophyte).

  b. A carpel (or two or more carpels fused together) has an ovary where the eggs develop, fertilization takes place, and seeds mature. A stigma is a sticky or hairy surface tissue above the lower portion of the ovary. It captures pollen grains and promotes their germination.

3. Inside carpels, ovules form on the inner ovary wall. Each consists of a female gametophyte with an egg cell, an endosperm mother cell, a surrounding tissue, and one or two protective layers called integuments. *31.3*

  a. At maturity, each ovule is a seed; its integuments form the seed coat.

  b. The ovary, and sometimes other tissues, matures into a fruit, which is a seed-containing structure.

4. Female gametophytes typically form this way: *31.3*

  a. Four haploid megaspores form after meiosis. All but one usually disintegrate.

  b. The one remaining megaspore undergoes mitosis three times but not cytoplasmic division. Its multiple nuclei move to prescribed positions in the cytoplasm.

  c. With cytoplasmic division, a female gametophyte (in this case, a seven-cell, eight-nuclei body) develops. One cell is an egg. The cell with two nuclei (endosperm mother cell) will help give rise to endosperm, which is a nutritive tissue for the forthcoming embryo.

5. Pollination refers to the arrival of pollen grains on a suitable stigma. Once it has been transferred, a pollen grain germinates. It becomes a pollen tube that grows down through tissues of the ovary, carrying the two sperm nuclei with it. *31.3*

6. At double fertilization, one sperm nucleus fuses with one egg nucleus, thereby forming a diploid (2*n*) zygote. The other sperm nucleus fuses with both nuclei of the endosperm mother cell to form a cell that will give rise to nutritive tissue in the seed. *31.3*

7. After double fertilization, the endosperm forms, the ovule expands, and the seed and fruit mature. *31.4*

8. Fruits function to protect and disperse seeds, which germinate after dispersal from the parent plant. As an example, fruits can attract animals and other dispersal agents; and lightweight fruits can be dispersed by wind. Some have hooks and such that attach to animals. *31.5*

9. Many flowering plants can reproduce asexually by vegetative growth. For example, new shoot systems may arise by mitotic divisions at nodes or buds along modified stems of a parent plant, by parthenogenesis or by vegetative propagation—from fragments or parts severed from a parent plant. *31.7*

Review Questions

1. Define flower and define pollinator. What kinds of animals are attracted to flowers with red and orange components? What kinds respond to the flowers that smell like decaying organic material? What are some of the ways in which night-foraging animals find flowers in the dark? *CI*

2. Label the floral parts. Explain the role each part plays in the reproduction of flowering plants. *31.1*

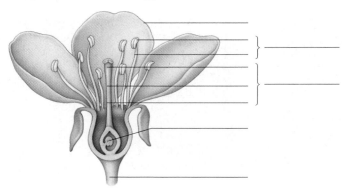

3. Distinguish between these terms:
  a. Sporophyte and gametophyte *31.1*
  b. Stamen and carpel *31.1*
  c. Ovule and ovary *31.1, 31.3*
  d. Microspore and megaspore *31.3*
  e. Pollination and fertilization *CI, 31.3*
  f. Pollen grain and pollen tube *31.1, 31.3*

4. Describe the steps by which an embryo sac, a type of female gametophyte, forms in a dicot such as cherry (*Prunus*). *31.3*

5. Define the difference between a seed and a fruit. *31.4*

6. Do food reserves accumulate in endosperm, cotyledons, or both? If both, in what ways do these structures differ? *31.4*

7. Name an example of a simple dry fruit, simple fleshy fruit, accessory fruit, aggregate fruit, and multiple fruit. *31.4*

8. Name the three regions of a fleshy fruit. What is the name for all three regions combined? *31.4*

9. Define and give an example of an asexual reproductive mode used by a flowering plant. *31.7*

Self-Quiz ANSWERS IN APPENDIX III

1. The flowers of many species coevolved with insects, birds, and other agents that function as _____ .

2. The _____ , which bears flowers, roots, stems, and leaves, dominates the life cycle of flowering plants.
   a. sporophyte
   b. gametophyte
   c. sporangium and its derivatives
   d. gametangium and its derivatives

3. Male gametophytes of flowering plants produce _____ , and the female gametophytes produce _____ .
   a. megaspores; eggs   c. eggs; sperm
   b. sperm; microspores   d. sperm; eggs

4. A _____ is a closed vessel that contains an ovary in which eggs develop, fertilization occurs, and seeds mature.
   a. pollen sac   c. receptacle
   b. carpel   d. sepal

5. Seeds are mature _____ ; fruits are mature _____ .
   a. ovaries; ovules   c. ovules; ovaries
   b. ovules; stamens   d. stamens; ovaries

6. After meiosis within pollen sacs, haploid _____ form.
   a. megaspores   c. stamens
   b. microspores   d. sporophytes

7. Following meiosis in ovules, _____ megaspores form.
   a. two   c. six
   b. four   d. eight

8. The seed coat forms from which structure(s)?
   a. ovule integuments   c. endosperm
   b. ovary   d. residues of sepals

9. Cotyledons develop as part of all flowering plant _____ .
   a. seeds   c. fruits
   b. embryos   d. ovaries

10. The outermost region of a fleshy fruit is the _____ .
    a. pericarp   c. mesocarp
    b. endocarp   d. exocarp

11. Development of a new plant from a tissue or structure that drops or is separated from the parent plant is called _____ .
    a. parthenogenesis
    b. exocytosis
    c. vegetative propagation
    d. nodal growth

12. Match the terms with the most suitable description.
    ____ double fertilization
    ____ ovule
    ____ mature female gametophyte
    ____ asexual reproduction
    ____ coevolution

    a. formation of zygote and first cell of endosperm
    b. outcome of two species interacting in close ecological fashion over geologic time
    c. contains a female gametophyte and has potential to be a seed
    d. an embryo sac, commonly with seven cells (one has two nuclei)
    e. mitotic cell division at bud or node produces a new plant

Critical Thinking

1. Wanting to impress her many friends with her sophisticated botanical knowledge, Dixie Bee is preparing a plate of tropical fruits for a party and cuts open a papaya (*Carica papaya*) for the first time. Inside are great numbers of slime-covered seeds, surrounded by soft flesh and soft skin (Figure 31.13). Knowing

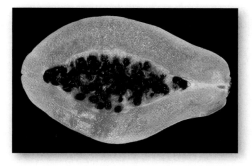

Figure 31.13 Seeds in the fruit of a papaya (*Carica papaya*).

that her friends will ask what sort of fruit this is and which parts are the endocarp, mesocarp, and exocarp, she panics, runs to her biology book, and opens it to Section 31.4. What does she find out? What will she tell them when they ask what kinds of agents disperse the seeds of such a fruit from the parent plant?

2. Observe several kinds of flowers growing in the area where you live. Given the coevolutionary links between flowering plants and their pollinators, describe what sorts of pollination agents your floral neighbors might depend upon.

3. Elaine, a plant physiologist, succeeded in cloning genes for pest resistance into petunia cells. How can she use tissue culture propagation to produce many petunia plants with those genes?

4. Before cherries, apples, peaches, and many other fruits ripen and the seeds inside them mature, their flesh is bitter or sour. Only later does it become tasty to animals that assist in seed dispersal. Develop a hypothesis of how this feature improves the odds for the plant's reproductive success.

Selected Key Terms

carpel 31.1	ovule 31.3
coevolution CI	parthenogenesis 31.7
cotyledon 31.4	pericarp 31.4
double fertilization 31.3	pollen grain 31.1
endosperm 31.3	pollination 31.3
flower CI	pollinator CI
fruit 31.4	seed 31.4
gametophyte 31.1	sporophyte 31.1
integument 31.3	stamen 31.1
megaspore 31.3	tissue culture propagation 31.7
microspore 31.3	vegetative growth 31.7
ovary 31.1	

Readings

Grant, M. October 1993. "The Trembling Giant." *Discover* 14(10): 82–89.

Proctor, M., and P. Yeo. 1973. *The Pollination of Flowers*. New York: Taplinger.

Raven, P., R. Evert, and S. Eichhorn. 1999. *Biology of Plants*. Sixth edition. New York: Freeman.

Rost, T., M. Barbour, C. R. Stocking, and T. Murphy. 1998. *Plant Biology*. Belmont, California: Wadsworth.

Stern, K. 2000. *Introductory Plant Biology*. Eighth edition. New York: McGraw-Hill.

32

PLANT GROWTH AND DEVELOPMENT

Foolish Seedlings, Gorgeous Grapes

A few years before the American stock market grew feverishly and then collapsed catastrophically in 1929, a researcher in Japan came across a substance that caused runaway growth and subsequent collapse of rice plants. Ewiti Kurosawa was studying what the Japanese call *bakane*, or the "foolish seedling" effect on rice plants. Stems of rice seedlings that had become infected with *Gibberella fujikuroi*, a fungus, grew twice as long as the stems of uninfected plants. The elongated stems were weak and spindly. Eventually they fell over, and the infected plants died. Kurosawa soon discovered that he could trigger the disease by applying extracts of the fungus to plants. Many years later, other researchers purified the disease-causing substance from fungal extracts. The substance was named gibberellin.

Gibberellin, as we now know, is one of the premier plant hormones. **Hormones**, remember, are signaling molecules secreted by some cells that travel to target cells, where they stimulate or inhibit gene activity (Section 15.5). Any cell that bears molecular receptors for a given hormone is its target. And the targets may be in the same tissue or some distance away.

Researchers have isolated more than eighty different forms of gibberellin from the seeds of flowering plants as well as from fungi. The changes that gibberellins trigger make young cells elongate, which makes stems lengthen (Figure 32.1). In nature, gibberellins help seeds and buds break dormancy and resume growth in spring.

Expose a cabbage plant to a suitable concentration of a gibberellin and its growth may well astound you (Figure 32.2). Applications of gibberellins make celery stalks longer and crispier, and they prevent the skin of navel oranges in orchard groves from ripening too quickly. Walk past those plump seedless grapes in the produce bins of grocery stores and observe how fleshy fruits of the grape plant (*Vitis*) grow in bunches along stems. Gibberellin applications made young cells elongate and stems lengthen between the internodes. This opened up more space between individual grapes, which thereby grew larger. Air could circulate better between grapes, which made it harder for pathogenic, grape-loving fungi to take hold and do damage.

Together, gibberellin and other plant hormones orchestrate plant growth and development. They also

Figure 32.1 Demonstrations of hormonal effects on plant growth and development. (**a**) Seedless grapes radiating market appeal. Gibberellin caused the stems to lengthen, which improved air circulation around grapes and gave them more room to grow. Grapes got larger and weighed more, making growers happy (grapes are sold by weight). (**b**) This young California poppy (*Eschscholzia californica*) was left alone. (**c**) Gibberellin was applied to this young poppy plant.

Figure 32.2 Foolish cabbages! To the left, in front of the ladder, are two untreated cabbages that were the controls. To the right, three cabbage plants treated with gibberellins

respond to cues from rhythmic changes of the seasons, as when days grow longer and warmer in spring after the short, cold nights of winter. Those grapes, cabbage leaves, or celery stalks that find their way into your mouth are the culmination of a plant's own exquisitely controlled programs of growth and development.

Here we continue with a story that started in the preceding chapter. That chapter traced the formation and development of a flowering plant zygote into a mature embryo, housed in a protective seed coat. At some point after its dispersal from a parent plant, the embryo is transformed into a seedling, which in turn grows and develops into a mature sporophyte. In time, the sporophyte typically forms flowers, then seeds and fruits. Depending on the species, it drops old leaves throughout the year or all at once, in autumn.

This part of the story surveys the heritable, internal mechanisms that govern plant development, and the environmental cues that turn such mechanisms on or off at different times, in different seasons.

Key Concepts

1. From the time a plant seed germinates, a number of hormones influence growth and development.

2. Hormones are signaling molecules between cells. One cell type produces and then secretes a particular kind of hormone, which stimulates or inhibits gene activity in other cell types that take up molecules of the hormone. The changes in gene activity have predictable effects, as when they trigger the mitotic cell divisions and other processes that make stems grow longer.

3. The known plant hormones are auxins, gibberellins, cytokinins, abscisic acid, and ethylene. We also have indirect evidence of other kinds of hormones, which have not yet been identified.

4. Plant hormones govern predictable patterns of development, including the extent and direction of cell growth and differentiation in particular plant parts.

5. Plant hormones help adjust patterns of growth and development in response to environmental rhythms, including seasonal shifts in daylength and temperature. In addition, they help adjust the patterns of response to environmental circumstances in which an individual plant finds itself, such as the amount of sunlight or shade, moisture, and so on at a given site.

6. Commonly, two or more kinds of plant hormones must interact with one another to bring about specific effects on growth and development.

PATTERNS OF EARLY GROWTH AND DEVELOPMENT—AN OVERVIEW

How Do Seeds Germinate?

Let's first consider the overall patterns of growth and development for dicots and monocots, starting with the events that take place inside a seed. Figure 32.3 shows the embryo sporophyte inside a grain of corn. (A grain, remember, is a seed-containing dry fruit.) The growth of the embryo idles before or after the seed is dispersed from the parent plant. Later, if all goes well, the seed germinates. **Germination** is the process by which some immature stage in the life cycle of a species resumes growth after a period of arrested development.

Germination depends on environmental factors, such as soil temperature, moisture, and oxygen level, and the number of daylight hours. The factors vary with the seasons. For instance, mature seeds don't hold enough water for cell expansion or metabolism. In many land habitats, ample water is available on a seasonal basis, so seed germination coincides with the spring rains. By the process of **imbibition**, water molecules move into a seed, being attracted mainly to hydrophilic groups of proteins stored in endosperm or cotyledons. As more water moves in, the seed swells and its coat ruptures.

Figure 32.4 (**a,b**) Pattern of growth and development of a dicot, the common bean plant (*Phaseolus vulgaris*). When this seed germinates, the embryo in it resumes growth. A hypocotyl (a hook-shaped part of the shoot below the cotyledons) forces a channel through soil, so the food-storing cotyledons are pulled up without being torn apart. At the surface, sunlight makes the hook straighten. Photosynthetic cells make food for several days in the cotyledons, which then wither and fall off. Foliage leaves take over photosynthesis. Flowers form in buds at nodes. (**c**) Cotyledons growing out of a seed coat.

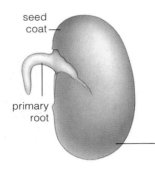

seed coat

primary root

a Bean seedling at the close of germination.

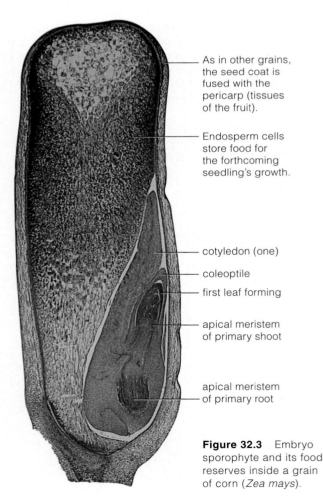

As in other grains, the seed coat is fused with the pericarp (tissues of the fruit).

Endosperm cells store food for the forthcoming seedling's growth.

cotyledon (one)

coleoptile

first leaf forming

apical meristem of primary shoot

apical meristem of primary root

Figure 32.3 Embryo sporophyte and its food reserves inside a grain of corn (*Zea mays*).

Figure 32.5 Pattern of growth and development of a monocot, corn (*Zea mays*). (**a,b**) A corn grain germinates. As the seedling grows through soil, a thin sheath (coleoptile) protects the new leaves. In corn seedlings, adventitious roots develop at the base of the coleoptile. (**c**) Coleoptile and primary root of a corn seedling. (**d**) The coleoptile and first foliage leaf of two seedlings poking out from the soil surface.

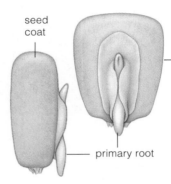

seed coat

primary root

a Corn grain at the close of germination.

Once the seed coat splits, more oxygen reaches the embryo, and aerobic respiration moves into high gear. The embryo's meristematic cells now divide rapidly. The root meristem is usually activated first. Its descendants divide, elongate, and give rise to the first, primary root of a seedling sporophyte. Germination is over when the seedling's primary root breaks through the seed coat.

Genetic Programs, Environmental Cues

Figures 32.4 and 32.5 show patterns of germination, growth, and development for dicots and monocots. The patterns have a heritable basis, being dictated by genes. All cells in a plant arise from the same cell (a zygote), so they generally inherit the same genes. But unequal divisions between daughter cells and their positions in

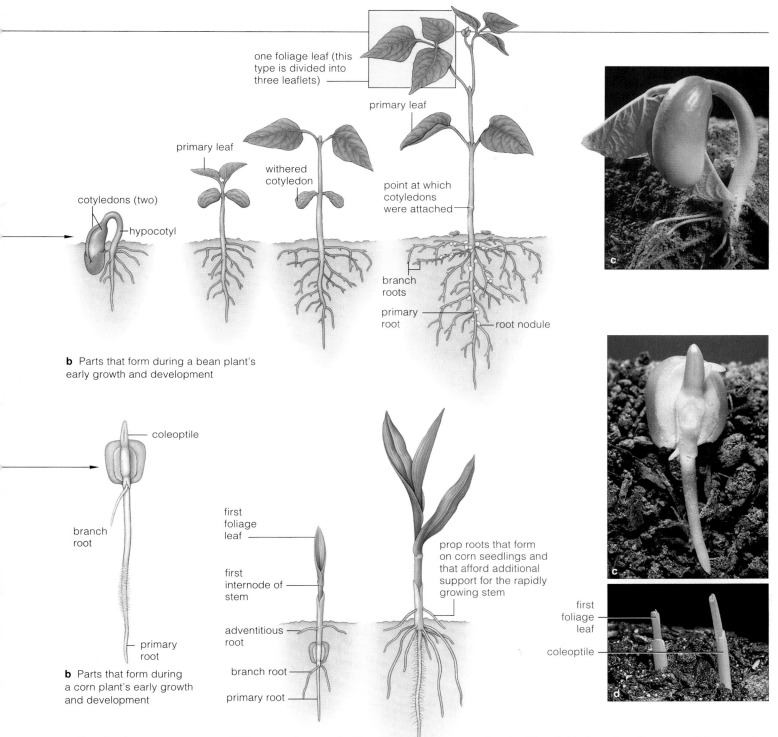

one foliage leaf (this type is divided into three leaflets)

primary leaf

primary leaf

withered cotyledon

point at which cotyledons were attached

cotyledons (two)

hypocotyl

branch roots

primary root

root nodule

c

b Parts that form during a bean plant's early growth and development

coleoptile

branch root

primary root

first foliage leaf

first internode of stem

prop roots that form on corn seedlings and that afford additional support for the rapidly growing stem

adventitious root

branch root

primary root

b Parts that form during a corn plant's early growth and development

c

first foliage leaf

coleoptile

d

a developing plant lead to differences in metabolic mechanisms and output. Cell activities start to differ with the onset of selective gene expression (Chapter 15). For example, the genes that govern the synthesis of growth-stimulating hormones are activated in some cells but not others. As you will see shortly, such events seal the developmental fate of different cell lineages.

Bear in mind, plants commonly adjust prescribed growth patterns in response to unusual environmental pressures. Suppose a seed germinates in a vacant lot and a heavy paper bag blows on top of it. The primary shoot will quickly bend and grow out from under the bag, in the direction of sunlight impinging on it. Interactions among enzymes, hormones, and other gene products in its cells result in this growth response.

Plant growth involves cell divisions and cell enlargements. Plant development requires cell differentiation, as brought about by selective gene expression.

Interactions among genes, hormones, and the environment govern how each individual plant grows and develops.

WHAT THE MAJOR PLANT HORMONES DO

When a plant undergoes **growth**, the number, size, and volume of its cells increase. The cells form specialized tissues, organs, and organ systems through successive stages during **development**. These cells communicate with one another by secreting signaling molecules into extracellular fluid and selectively responding to signals from other cells. Hormones are foremost among the signals. As you know from Chapters 15 and 28, after a signal is received by a target cell, it gets transduced in a way that causes a specific change in metabolism or gene expression. Signal transduction often involves changes in the shape of the receptor, then activation of enzymes and of second messengers in the cytoplasm.

Growth proceeds by cell divisions and enlargements at meristems, as described in Chapter 29. About half of the resulting daughter cells do not increase in size, but they retain their capacity to divide. Their descendants divide in genetically prescribed planes and expand in prescribed directions to give rise to parts with specific shapes and functions. Figure 32.6 shows an example.

Young cells enlarge as they take up water and fluid pressure, or *turgor* pressure, acts on their primary wall. Just as a soft balloon inflates easily, a soft-walled cell expands under turgor pressure. Its wall thickens with polysaccharide additions, and more cytoplasm forms.

Hormones affect the selective gene expression that underlies patterns of plant growth and development, starting with seed germination. Table 32.1 lists the five

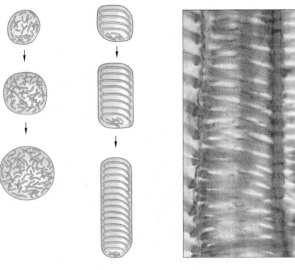

a Growth proceeds in all directions

b Growth in longitudinal direction only

c Cellulose rings thicken the secondary wall of this tracheid, a conducting tube of xylem

Figure 32.6 How cells get their shape. Microtubules in the cytoplasm inherited from a parent cell are already arranged in patterns that govern how cellulose microfibrils will be oriented in the cell wall. (**a**) On exposure to ethylene, their orientation becomes more random. Because a primary wall is elastic all over, the cell expands in all directions and assumes a spherical shape. (**b**,**c**) On exposure to gibberelins, microtubules assume a transverse orientation and the cellulose microfibrils become similarly arranged. Such a cell can only lengthen.

Table 32.1 *Overview of Plant Signaling Molecules and Their Effects*

Signaling Molecule	Source and Mode of Transport	Stimulatory or Inhibitory Effects
HORMONES		
Gibberellins	Synthesized in apical meristem of bud, leaf, roots, seeds. May travel in xylem and phloem	Makes stems lengthen by stimulating cell division and elongation; often contributes to flowering; helps end bud, seed dormancy
Auxins	Synthesized mainly in apical meristem of bud, leaf, seed. Diffuses from cells in one direction	Stimulates apical dominance, tropisms, vascular cambium divisions, vascular tissue development, fruit formation. Inhibits leaf, fruit abscission
Cytokinins	Synthesized mainly in root tip. Travels from roots to shoots in xylem	Stimulates cell division, leaf expansion. Inhibitory effect on leaf aging. Its application can release buds from apical dominance
Ethylene (a gas)	Synthesized in most parts undergoing ripening, aging, or stress. Diffusion in all directions	Stimulates fruit ripening, abscission of leaf, flower, fruit
Abscisic acid (ABA)	Synthesized in mature leaf in response to water stress; also in stems, unripened fruit	Stimulates stomatal closure, development of embryo sporophytes, distribution of photosynthetic products to seeds, product storage and protein synthesis in seeds. May influence dormancy in some species
GROWTH REGULATORS		
Brassinolides	Steroid	Influences cell division, elongation; required for normal growth; protects against pathogens, stress
Jasmonates	Volatile compound derived from fatty acid	Influences seed germination, root growth, protein storage, defense
Salicylic acid	Phenolic compound structurally like aspirin	Influences tissue defense responses to pathogens
Systemin	Peptide; synthesized in damaged tissues	Influences defense responses in damaged tissues

Figure 32.7 Examples of the effects of auxin. (**a**) *Left:* A cutting from a gardenia plant four weeks after auxin was applied to its base. *Right:* An untreated cutting used as a control.

(**b**) Experiments demonstrating how IAA in a coleoptile tip causes elongation of cells below it. (1) First, cut off the tip of an oat coleoptile. The cut stump will not elongate as much as a normal oat coleoptile (2) used as the control. (3) Place a tiny block of agar under the cut tip and leave it for several hours. IAA diffuses into the agar. (4) Now put the agar on top of another de-tipped coleoptile. Its elongation will proceed about as rapidly as in an intact coleoptile growing next to it (5).

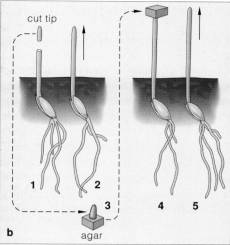

major classes of hormones. They are the gibberellins, auxins, cytokinins, abscisic acid, and ethylene.

Gibberellins stimulate cell division and elongation. As described earlier, they make stems lengthen. They help seeds break dormancy in spring by making cells of the primary root lengthen. In at least some cases, they influence flowering and the development of fruit.

Auxins control cell elongations that lengthen shoots and coleoptiles. A **coleoptile** is a thin sheath that keeps some primary shoots from shredding as growth pushes them up through soil (Figure 32.7*a*). In addition, auxins help vascular tissues and vascular cambium develop. They help keep leaves, flowers, and fruits from falling prematurely. Also, auxin produced in seeds stimulates growth of the ovary wall during fruit formation.

Auxins form in the apical meristems of shoots and coleoptiles. They travel a long distance toward the base of the plant by diffusing through parenchyma cells in vascular tissues. Auxin also causes *apical dominance*, an inhibition of lateral bud growth. Gardeners routinely pinch off the shoot tips to stop auxin from diffusing through stems, so lateral buds are free to branch out.

IAA (indoleacetic acid) is the most pervasive auxin in nature. Orchardists use it to thin flowers in spring so trees yield fewer but larger fruits. They also use it to prevent premature fruit drop so all fruit in the orchard can be picked at the same time, which cuts labor costs.

2,4-D, a synthetic auxin, is a widely used **herbicide** (Section 32.6). It does not seem to harm humans when properly handled. But mixing it with equal parts of a related compound, 2,4,5-T, yields *Agent Orange*. This herbicide was used to clear forested war zones during the Vietnam conflict. Later, experiments linked dioxin, a carcinogen and trace contaminant of 2,4,5-T, to birth defects, miscarriages, leukemias, and liver disorders. 2,4,5-T is now banned in the United States.

Cytokinins stimulate rapid cell division (the name refers to cytoplasmic division). They are abundant in root and shoot meristems and in maturing fruits. They oppose auxin's effects by promoting growth of lateral buds. At least in tissue cultures, cytokinins and auxins work together in promoting rapid cell divisions. Also, cytokinins keep leaves from aging before their time. They are used for basic research and to prolong the shelf life of cut flowers and other horticultural goods.

Abscisic acid (ABA) helps plants adapt to seasonal changes, as by inducing bud dormancy and inhibiting cell growth and premature seed germination. ABA also contributes to stomatal closure when a plant is water stressed (Section 30.4). Growers typically apply ABA to nursery stock before shipping it, because plants are not as vulnerable to injury when they are dormant.

Ethylene, the only gaseous plant hormone, induces fruit ripening, leaf drop, and other aging responses. Ancient Chinese knew to burn incense to ripen fruit faster. By the early 1900s, growers were ripening citrus fruits by storing them in sheds with kerosene stoves. Food distributors now use ethylene to ripen tomatoes and other green fruit after they ship them to grocery stores. Fruit picked green doesn't bruise or deteriorate as fast. Ethylene exposure brightens citrus rinds before the fruit is displayed in the market.

Besides the hormones just described, plants also produce growth regulators of the sort listed in Table 32.1. There are more than these, including some as-yet unknown signal or signals that induce flowering.

Plant hormones and growth regulators are required for normal plant growth and development.

Hormones are signaling molecules secreted by specific cells that influence gene activity in target cells, which may be nearby or some distance away in the multicelled body.

The main categories of plant hormones are gibberellins, auxins, cytokinins, abscisic acid, and ethylene.

ADJUSTING THE DIRECTION AND RATES OF GROWTH

What Are Tropisms?

Generally, the young roots of land plants grow down through soil, and the shoots grow upright through air. Both also can adjust the direction of growth in response to environmental stimuli, as when a new shoot turns toward sunlight. When any root or shoot turns toward or away from an environmental stimulus, we call this a plant tropism (after the Greek *trope*, for turning). As the following examples illustrate, these responses are an outcome of hormone-mediated shifts in the rates at which different cells in the plant grow and elongate.

GRAVITROPISM The first root to break through a seed coat always curves downward, and the coleoptiles and new stems curve upward. Figure 32.8*a* has an example. These are growth responses to the Earth's gravitational force; they are forms of **gravitropism**.

Figure 32.8*b* shows how to track this response for a seedling turned on its side in a dark room. The stem curves up even in the absence of light. How? In that horizontally oriented stem, cell elongations slow down

greatly on the side facing up, but elongations on the lower side increase fast. Different rates of elongation on opposing sides of the stem are enough to make it bend upward. Something apparently makes cells that are on the "bottom" of a stem turned on its side *more* sensitive to a hormone, and those on top *less* so.

Auxin, together with a growth-inhibiting hormone in roots, may trigger such responses. Turn a young root on its side and remove its root cap, and it will *not* curve down. Put the cap back on, and the root curves down. Elongating root cells won't stop growing if you remove the root cap; if anything, cells will grow faster. Suppose a growth inhibitor in root cap cells gets *redistributed* inside a root turned on its side. If gravity causes the inhibitor to leave the cap and accumulate in cells on the root's lower side, the cells won't elongate as much as cells on the upper side. The root will curve down.

Gravity-sensing mechanisms of plants are based on **statoliths**, which generally are clusters of particles in a number of different cells. In plants, they are clusters of unbound starch grains in modified plastids. Figure 32.9 gives one example. These plastids are made denser by the statoliths. They collect near the bottom of root cells in response to gravity and settle downward until they rest in the lowest cytoplasmic region. Statolith redistribution may cause an *auxin* redistribution inside the cells, which initiates the gravitropic response.

PHOTOTROPISM When stems or leaves adjust the rate and direction of growth in response to light, they are showing **phototropism**. These adjustments favor the light-dependent reactions of photosynthesis, which do not proceed in the dark. Plants that reorient themselves to maximize light interception have an advantage.

Charles Darwin pondered on phototropism after it dawned on him that a coleoptile was growing toward light on one side of its tip. But it wasn't until the 1920s that Fritz Went, a graduate student in Holland, linked phototropism with a growth-promoting substance. He was the person who named the substance auxin (after the Greek *auxein*, "to increase"). Went demonstrated that auxin moves from the tip of a coleoptile into cells

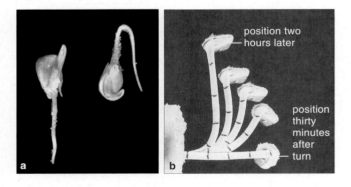

Figure 32.8 Observational tests of gravitropic responses by plants. (**a**) Responses of a corn primary root and shoot that are growing in their normal orientation (*left*) and when turned upside down (*right*). (**b**) One way to measure a gravitropic response. Force a sunflower seedling to grow in a darkened room for five days. Then turn it on its side, mark the shoot at 0.5-centimeter intervals, and watch what happens.

Figure 32.9 Observational test of the gravitropic responses by young roots. (**a**) Normal orientation of plastids in root cap cells of corn. Statoliths in these cells settle downward. (**b**) Turn the root sideways. Five to ten minutes later, plastids have settled to the new "bottom" of the cells. A gravity-sensing mechanism using statoliths may be sensitive to auxin redistribution in a root tip. A difference in auxin concentrations may cause cells on the "top" of a root turned sideways to elongate faster than cells on the bottom. Different elongation rates will make the root tip curve downward.

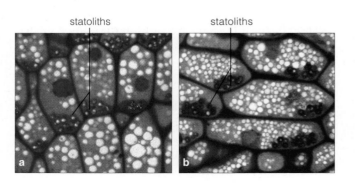

b Rays of sunlight strike one side of a coleoptile.

c Coleoptile bends after auxin diffuses from its tip to cells on its shaded side.

Figure 32.10 (**a**) Phototropism by tomato seedlings. (**b,c**) Hormone-mediated differences in the rates of cell elongation induce coleoptiles and stems to bend toward light.

less exposed to light and makes them elongate faster than cells on the illuminated side. The difference in rates of growth results in bending toward light.

You can observe a phototropic response by putting seedlings of sun-loving plants in a dark room next to a window through which rays of sunlight are streaming. Tomato seedlings will do. They will start curving in the direction of the most light (Figure 32.10).

Blue wavelengths of light are known to induce the strongest phototropic response. These wavelengths can be absorbed by **flavoprotein**. This is a yellow pigment, and it may be the receptor that transduces light energy into the phototropic bending mechanism.

THIGMOTROPISM Plants also shift their direction of growth when they contact solid objects. This response is called **thigmotropism** (after the Greek *thigma*, which means "touch"). Auxin and ethylene may have roles in the response. Vines, which are stems too slender or soft to grow upright without support, make such a contact response. So do tendrils, which are modified leaves and stems that wrap around objects and help support the plant (Figure 32.11). You can observe what happens as they grow against a stem of another plant. Within minutes, cells on the contact side stop elongating and the vine or tendril starts curling around the stem, often more than once. Afterward, the cells on both sides will resume growth at the same rate.

Responses to Mechanical Stress

Mechanical stress, as inflicted by prevailing winds and grazing animals, can inhibit stem elongation and plant growth. You can see such effects on trees growing near the snowline of windswept mountains; they are stubby compared to trees of the same species that are growing at lower elevations. Similarly, plants grown outdoors commonly have shorter stems than plants grown in a

Figure 32.11 Passion flower (*Passiflora*) tendril busily twisting thigmotropically.

Figure 32.12 Effect of mechanical stress on tomato plants. (**a**) This plant, the control, grew in a greenhouse. (**b**) Each day for twenty-eight days, this plant was mechanically shaken for thirty seconds. (**c**) This one had two shakings each day.

greenhouse. You can observe this response to stress by shaking a plant daily for a brief period. Doing so will inhibit growth of the whole plant (Figure 32.12).

Plants adjust the direction and rate of their growth in response to environmental stimuli.

HOW DO PLANTS KNOW WHEN TO FLOWER?

All flowers are variations on the same pattern of growth and development. In Section 28.5, you read about the expression of genes that control floral development in *Arabidopsis thaliana*. A hormonal signal activates these genes. But what activates the hormone-secreting cells?

An Alarm Button Called Phytochrome

All organisms have internal mechanisms that preset the time for recurring changes in biochemical events. Some of the internal timing mechanisms—**biological clocks**—trigger shifts in daily activities. Section 28.5 gave one example, the rhythmic leaf movements of a bean plant. That leaf movement is a *circadian* rhythm, a biological activity repeated in cycles, each lasting for close to twenty-four hours. Experiments by Ruth Satter, Richard Crain, and their colleagues at the University of Connecticut showed that phytochrome has a role in leaf movements. Satter, a pioneer in the study of timing mechanisms, was one of the first to correlate plant rhythms with "hands of a biological clock."

Biological clocks also help bring about seasonal adjustments in basic patterns of growth, development, and reproduction—including flower formation.

Some biological clocks have an alarm button called **phytochrome**. This blue-green pigment is part of the switching mechanisms that promote or inhibit growth for a variety of plant parts. Phytochrome is a receptor for red and far-red wavelengths of light. At sunrise, red wavelengths dominate the sky. They are the signal that causes phytochrome to change shape to Pfr, its active molecular form (Figure 32.13). The signal is transduced in one of two ways. Either cells are induced to take up free calcium ions (Ca^{++}) or organelles are induced to

release them. The response starts as ions combine with calcium-binding proteins in cells. At sunset, at night, or in the shade, the signal transduction pathway is reversed. At such times, wavelengths are primarily far-red. They cause phytochrome to revert to its inactive molecular form (Pr).

Photoperiodism is a biological response to change in the length of daylight relative to darkness in a cycle of twenty-four hours. Phytochrome's active form, Pfr, sets in motion events that result in the transcription of genes. The gene products include signaling molecules that deal with seed germination, shoot elongation and branching, expansion of leaves, flower, fruit, and seed formation, and dormancy (Figure 32.14).

Flowering—A Case of Photoperiodism

Different flowering plants start diverting more energy to forming flowers at different times of year. Tulips flower in spring, for example, and chrysanthemums in autumn. It seems probable that the secretion of one or more flower-inhibiting and flower-inducing hormones depends on timed phytochrome responses to cues from the environment. Despite intensive searches, however, no one has yet identified the elusive hormone(s).

It is a puzzle. Tulips, spinach, and other **long-day** plants flower in the spring, when daylength exceeds some critical value. Chrysanthemums, poinsettias, and cockleburs are among the **short-day** plants. They flower in late summer or early autumn, when daylength is shorter than a critical value. **Day-neutral** plants simply flower when they are mature enough to do so.

Actually, these are not very good names, because the environmental cue is *night length,* not daylength.

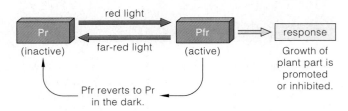

Figure 32.13 Interconversion of the phytochrome molecule from active form (Pfr) to inactive form (Pr). This blue-green pigment is part of a switching mechanism that promotes or inhibits the growth of a variety of plant parts.

Figure 32.14 Plant growth and development as correlated with the number of hours of light available each day. The number changes with the passing of the seasons. The data shown reflect photoperiodic responses of plants that grow in temperate regions of North America. In such regions, rainfall and temperature shift with the seasons.

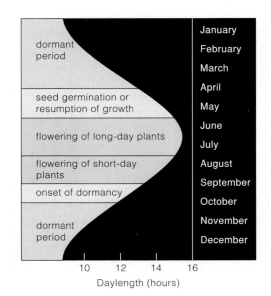

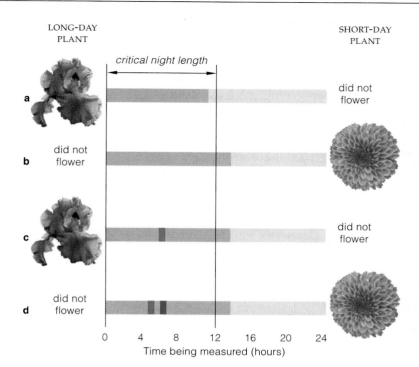

Figure 32.15 Experiments showing that short-day plants flower by measuring night length. Each horizontal bar signifies 24 hours. *Yellow* bars signify daylight; *blue* bars signify night. (**a**) Long-day plants flower when the night is *shorter* than a length that is critical for flowering. (**b**) Short-day plants flower when the night is *longer* than a critical value. (**c**) When an intense red flash interrupts a long night, both kinds of plants respond as if it were a short night. (**d**) A short pulse of far-red light after the red flash cancels the disruptive effect of the red flash.

Figure 32.16 Experiment involving flowering responses of (**a**) spinach, a long-day plant, and (**b**) chrysanthemum, a short-day plant. In both photographs, the plant at *left* grew under short-day conditions. The plant at *right* grew under long-day conditions.

Figures 32.15 and 32.16 show the responses of long-day and short-day plants to a range of light conditions. They all respond to the wavelengths that predominate at dawn and at dusk. When experimenters interrupt a critical dark period with a pulse of red light, they reset the clocks. "Short-day" plants flower only when nights are longer than a critical value. And "long-day" plants flower when the night is shorter than a critical value.

And so spinach plants won't flower and form seeds unless they are exposed to ten hours of darkness for two weeks. That's why you would not want to start, say, a spinach seed farm in the tropics, where that set of cues is not available. Chrysanthemum growers stall the flowering process by exposing plants to a flash of light at night to break up one long night into two short ones. That is why they can offer chrysanthemums in spring as well as autumn. Cockleburs normally flower after only one night longer than 8–1/2 hours. When artificial light interrupts their dark period for even a minute, they won't flower. And why didn't poinsettias planted along California's interstate highways put out flowers? Headlights from vehicles zipping past during the night inhibited the flowering response.

The unidentified hormone(s) that are thought to be central to flowering might be produced in leaf cells and transported to new floral buds. Experiments suggest this is so. For example, trim all but one leaf away from a cocklebur plant. Next, cover that leaf entirely with black paper for 8–1/2 hours. The plant will proceed to produce flowers. However, cut off that one cocklebur leaf right after the dark period is completed, and the plant will not produce flowers.

Like other organisms, flowering plants have biological clocks, which are internal time-keeping mechanisms.

Phytochrome, a blue-green pigment, is part of a switching mechanism for phototrophic responses to light of red and far-red wavelengths. Its active form, Pfr, might trigger the secretion of one or more hormones that induce and inhibit flowering at different times of year.

The main environmental cue for flowering is the length of night—that is, the hours of darkness, which vary with the seasons. Different kinds of plants form flowers at different times of year, depending on their phytochrome mechanism.

LIFE CYCLES END, AND TURN AGAIN

Senescence

While leaves and fruits are growing, cells inside them produce auxin (IAA), which moves into stems. There it interacts with cytokinins and gibberellins to maintain growth. When autumn approaches and the length of daylight decreases, plants start to withdraw nutrients from leaves, stems, and roots, then distribute them to flowers, fruits, and seeds. Deciduous plants shed leaves as a growing season ends. They channel nutrients to storage sites in twigs, stems, and roots before their leaves die and drop. The dropping of flowers, leaves, fruits, and other plant parts is known as **abscission**. An abscission zone is made of thin-walled parenchyma cells at the base of a petiole or some other part about to drop from the plant. Figure 32.17 shows one example.

Senescence is the sum total of processes that lead to death of a plant or some of its parts. The recurring cue is a decrease in daylength. But other factors, such as drought, wounds, and nutrient deficiencies, also bring it about. The cue causes a decline in IAA production in leaves and fruits. A different signal that the plant itself may produce stimulates abscission zone cells to make

ethylene. Cells enlarge, deposit suberin in their walls, and produce enzymes that digest cellulose and pectin in middle lamellae. A middle lamella is the cementing layer between plant cell walls (Section 4.11). While cells continue to enlarge and their walls become digested, they separate from one another. Eventually the leaf or other plant part above the abscission zone drops away.

Interrupt the diversion of nutrients into flowers, seeds, or fruits, and you stop aging of a plant's leaves, stems, and roots. For example, if you remove each new flower or seed pod from a plant, its leaves and stems will remain vigorous and green much longer (Figure 32.18). Gardeners routinely remove flower buds from many kinds of plants to maintain vegetative growth.

Entering Dormancy

As autumn approaches and days grow shorter, many perennials and biennials start to shut down growth. They do so even when temperatures are still mild, the sky is bright, and water is plentiful. When a plant stops growing under conditions that seem (to us) suitable for growth, it has entered a state of **dormancy** in which its metabolic activities idle. Ordinarily, the plant's buds will not resume growth until there is a convergence of precise environmental cues in early spring.

Short days, long, cold nights, and dry soil that is deficient in nitrogen are strong cues for dormancy. Researchers tested this by interrupting the long dark period of some Douglas firs with a short period of red light. The plants responded as if nights were shorter and days longer. They continued to grow taller (Figure 32.19). In this experiment, conversion of Pr to Pfr by red light during the dark period prevented dormancy.

tissues of stem   cells of abscission zone

Figure 32.17 Light micrograph of an abscission zone in a red maple (*Acer*), in longitudinal section. The zone forms at the base of leaf petioles.

control (pods not removed)   experimental plant (pods removed)

Figure 32.18 The observable results from an experiment in which seed pods were removed from a soybean plant. Removal delayed its senescence.

Figure 32.19 Experiment to test the effect of the relative length of day and night on Douglas firs. The fir at left was exposed to 12-hour light/12-hour dark cycles for one year. Its buds stayed dormant; "daylength" was too short. The fir at right was exposed to 20-hour light/4-hour dark cycles. It continued growing. The fir in the middle was exposed to 12-hour light/11-hour dark cycles with 1 hour of light interrupting the middle of the dark period. The interruption prevented bud dormancy; it caused Pfr formation at a sensitive time in the normal day–night cycle.

Potted plant grown inside a greenhouse did not flower.

Branch exposed to cold outside air flowered.

Figure 32.20 Localized effect of cold temperature on dormant buds of a lilac (*Syringa*).

For this experiment, one branch of the lilac plant was positioned so that it protruded from a greenhouse during a cold winter. The rest of the plant was kept inside and exposed only to warmer temperatures. Only buds on the branch exposed to low outside temperatures resumed growth in spring.

In nature, buds might enter dormancy because less Pfr can form when daylength shortens in late summer.

The requirement for multiple cues for dormancy has adaptive value. For example, if temperature were the only cue, then warm autumn weather might make plants flower and seeds germinate—and winter frost would kill them. By contrast, with artificial selection, growers have developed seeds that germinate promptly in greenhouses any time of year.

Breaking Dormancy

Temperatures as well as daylength change seasonally in most places, and dormancy-breaking mechanisms work between fall and spring. The temperatures often become milder, and rains and nutrients again become available. With the return of favorable conditions, life cycles turn once more. Seeds germinate; buds resume growth and give rise to new leaves, then to flowers.

Breaking dormancy probably requires gibberellins and abscisic acid. Often it does not occur without prior exposure to low temperatures at specific times of year. Temperatures required to break dormancy vary among plants. For example, Delicious apple trees growing in Utah require 1,230 hours near 43°F (6°C). Apricot trees grown there require 720 hours. Generally, trees in the southern United States require less cold exposure than the trees growing in northern states and Canada. If you

live in Colorado and order a young peach tree from a Georgia nursery, the tree might start spring growth too soon and be killed by late frost or heavy snow.

Vernalization

Flowering, too, is often a response to seasonal changes in temperatures. To give one example, unless buds of some biennials and perennials become exposed to low winter temperatures, flowers will not form on their stems in spring. The low-temperature stimulation of flowering is called **vernalization** (from *vernalis*, which means "to make springlike"). Figure 32.20 shows how you can gather experimental evidence of this effect.

As long ago as 1915, the plant physiologist Gustav Gassner studied flowering of some cereal plants after exposing their seeds to controlled temperatures. As an example, he germinated seeds of winter rye (*Secale cereale*) at near-freezing temperatures. Plants flowered the next summer even when they were planted in late spring. Vernalization is common in agriculture.

Multiple cues from the environment influence hormonal secretions that stimulate or inhibit processes of growth and development during the life cycle of plants. These cues include changes in daylength, temperature, moisture, and nutrient availability.

GROWING CROPS AND A CHEMICAL ARMS RACE

In this unit, you glimpsed the nature of plant structure and function. Think about that the next time you eat strawberries, corn, beans, or anything else grown on fertilized cropland. How did the plants you eat for nutrients get enough of those nutrients? How did they manage to compete with weeds and avoid pests that ruin or gobble up nearly half of what we try to grow?

Most plants aren't entirely defenseless. They evolved under selection pressure of attacks by insects and other organisms and often repel attackers with natural toxins. A **toxin** is an organic compound, a normal metabolic product of one species, but its chemical effects harm or kill a different organism that contacts it. Humans, too, encounter traces of natural toxins, even in such familiar foods as hot peppers, potatoes, figs, celery, rhubarb, and alfalfa sprouts. Still, we do not die in droves, so we must have chemical defenses against these toxins.

Figure 32.22 One of the crop dusters that intervene in the competition for nutrients between crop plants and pests, including weeds.

Just a few thousand years ago, farmers used sulfur, lead, arsenic, and mercury to protect crop plants from insects. They freely dispensed the highly toxic metals until the late 1920s, when someone figured out they were poisoning people. Traces of the toxic metals still turn up in contaminated lands.

The farmers also used organic compounds extracted from leaves, flowers, and roots as natural pesticides. In 1945, scientists started to make synthetic toxins and to identify mechanisms by which toxins attack pests. The *herbicides*, including the synthetic auxin mentioned in Figure 32.21a, kill weeds by disrupting metabolism and growth. *Insecticides* clog the airways of a target insect, disrupt its nerves and muscles, or prevent its reproduction. *Fungicides* work against harmful fungi, including the mold that makes aflatoxin, one of the deadliest poisons. By 1995, people in the United States were spraying or spreading more than 1.25 billion pounds of toxins through fields, gardens, homes, and industrial and commercial sites (Figure 32.22).

Some pesticides also kill birds and other predators that help control pest population sizes. Besides this, targeted pests have been developing resistance to the chemical arsenal, for reasons given earlier (Chapter 1). Pesticides cannot be released haphazardly, for they can be inhaled, ingested with food, or absorbed through skin. Different types are active for weeks or years. Some trigger rashes, headaches, hives, asthma, and joint pain in millions of people. Some trigger life-threatening allergic reactions in abnormally sensitive people.

DDT and other long-lasting pesticides are currently banned in the United States. Even rapidly degradable ones are subjected to rigorous application standards and ongoing safety tests.

And how we protect crop plants is just one aspect of what must be done to support a human population that now exceeds 6.1 billion. Should essential crop plants be genetically engineered? Can we, and should we, make them better at fixing nitrogen or tolerating salts? What about bruise resistance? Built-in pesticide resistance? Here you might reflect once more on Sections 28.2, 16.7, and the first *Critical Thinking* question in 16.11.

a *2,4-D* (2,4-dichlorophenoxyacetic acid), a synthetic auxin widely used as a herbicide. Enzymes of weeds and microbes cannot easily degrade it, compared to natural auxins.

b *Atrazine*, the bestselling herbicide, kills weeds weeds in a few days, as do glyphosate (Roundup), alachlor, (Lasso), and daminozide (Alar). It now appears that atrazine causes abnormal sexual development in frogs, even in trace amounts below the level allowed in drinking water.

c Dichlorodiphenyltrichloroethane, or *DDT*. It takes two to fifteen years for this nerve cell poison to break down. Chlordane, another type of insecticide, also persists for a long time in the environment.

d *Malathion*. Like other organophosphates, it is cheap, breaks down faster than chlorinated hydrocarbons, and is more toxic. Organophosphates represent half of all insecticides used in the United States. Some are now banned for crops; application of others must end at least three weeks before harvest. Farmers who contest this policy want the Environmental Protection Agency to consider economic and trade issues as well as human health.

Figure 32.21 A few pesticides, some more toxic than others.

How well plants grow and develop ultimately affects the growth and development of the human population.

SUMMARY 　　　　　　　　*Gold* indicates text section

1. This chapter started with what happens after seeds are dispersed from parent plants. Inside each seed, the embryo sporophyte has been dormant. It germinates by imbibition—it absorbs water, resumes growth, and breaks through its seed coat. Now control mechanisms govern the growth and development of the new plant. The seedling increases in volume and mass. Its tissues and organs develop. Later, fruits and new seeds form, then older leaves drop. *32.1*

2. How individual plants grow and develop depends on interactions among their genes, their hormones, and cues from the environment. *32.2*

 a. A plant's genes govern the synthesis of enzymes and other proteins necessary for metabolism, hence for all cell activities. How and when each type of enzyme functions depend on hormonal action.

 b. Hormones are a category of signaling molecules. After being produced and secreted by some cells, they travel to target cells in different parts of the plant body and stimulate or inhibit gene activity. As is the case for other kinds of organisms, any cell that bears receptors for a particular hormone is its target.

 c. The prescribed growth patterns are influenced by environmental cues. Often, they become adjusted in response to unusual environmental pressures.

3. Plant hormones bring about predictable patterns of growth and development. They also trigger responses to environmental rhythms (such as seasonal changes in daylength and temperature) and variations in shade, sunlight, and other circumstances. *32.1, 32.2*

4. Five major categories of plant hormones have been identified. Probably there are others. *32.2–32.5*

 a. Gibberellins promote stem elongation, they help seeds and buds break dormancy in spring, and they may help induce the flowering process.

 b. Auxins promote coleoptile and stem elongation. They have roles in phototropism and gravitropism.

 c. Cytokinins stimulate cell division, promote leaf expansion, and retard leaf aging.

 d. Abscisic acid promotes bud and seed dormancy, and it limits water loss by promoting stomatal closure.

 e. Ethylene promotes fruit ripening and abscission.

5. Plant parts make tropic responses to light, gravity, and other environmental conditions. Hormones induce a difference in the rate and direction of growth on two sides of the part, which causes it to turn or move. *32.3*

 a. With gravitropism, roots grow downward and stems grow upright in response to the Earth's gravity. Some gravity-sensing mechanisms in plants are based on statoliths (clusters of particles in cells).

 b. With phototropism, stems and leaves adjust rates and directions of growth in response to light. A yellow pigment (flavoprotein) might be involved; it absorbs blue wavelengths that trigger the strongest response.

 c. With thigmotropism, plants adjust their direction of growth in response to contact with solid objects.

6. Plants respond to mechanical stress, as when strong winds inhibit stem elongation and plant growth. *32.3*

7. A biological clock is any internal, time-measuring mechanism that has a biochemical basis. *32.4*

 a. Circadian rhythms are biological activities that recur in cycles that each last about twenty-four hours. Rhythmic movements of leaves are an example.

 b. Photoperiodism is a biological response to change in the relative length of daylight and darkness in the twenty-four-hour cycle. Photoperiodism is seasonal in plants. Phytochrome, a blue-green pigment, is part of a switching mechanism for a clock. It helps promote or inhibit germination, stem elongation, leaf expansion, stem branching, and flower, fruit, and seed formation.

 c. Long-day plants flower during spring or summer, when there are more hours of daylight than darkness. Short-day plants flower when daylength is less. Day-neutral plants flower regardless of daylength.

8. Senescence is the sum of processes leading to the death of a plant or plant structure. *32.5*

9. Dormancy is a state in which a perennial or biennial stops growing even when conditions appear suitable for continued growth. A decrease in Pfr levels might trigger dormancy. Breaking dormancy might involve exposure to certain temperatures and hormonal action, including gibberellins and abscisic acid. *32.5*

Review Questions

1. Explain the process by which seeds germinate. *32.1*

2. Briefly describe how cells of a new plant grow, enlarge, and take on specific shapes. *32.1*

3. List the five known types of plant hormones and describe the known functions of each. *32.2*

4. List a few plant growth regulators and their functions. *32.2*

5. Define plant tropism. What is the difference between phototropism and photoperiodism? *32.3, 32.4*

6. What is phytochrome, and what role does it play in the flowering process? *32.4*

7. Explain the differences between long-day, short-day, and day-neutral plants. *32.4*

8. Define dormancy and senescence. Give examples. *32.5*

Self-Quiz 　ANSWERS IN APPENDIX III

1. Seed germination is over when the _____ .
 a. embryo sporophyte absorbs water
 b. embryo sporophyte resumes growth
 c. primary root pokes out of the seed coat
 d. cotyledons unfurl

Figure 32.23 Field of sunflowers (*Helianthus*) that are busily demonstrating solar tracking.

2. Which of the following statements is false?
   a. Auxins and gibberellins promote stem elongation.
   b. Cytokinins promote cell division but retard leaf aging.
   c. Abscisic acid promotes water loss and dormancy.
   d. Ethylene promotes fruit ripening and abscission.

3. Plant hormones _____ .
   a. interact with one another
   b. are influenced by environmental cues
   c. are active in plant embryos within seeds
   d. are active in adult plants
   e. all of the above

4. Plant growth depends on _____ .
   a. cell division   c. hormones
   b. cell enlargement   d. all of the above

5. _____ are the strongest stimulus for phototropism.
   a. Red wavelengths   c. Green wavelengths
   b. Far-red wavelengths   d. Blue wavelengths

6. Light of _____ wavelengths makes phytochrome switch from inactive to active form; light of _____ wavelengths has the opposite effect.
   a. red; far-red   c. far-red; red
   b. red; blue   d. far-red; blue

7. The flowering process is a _____ response.
   a. phototropic   c. photoperiodic
   b. gravitropic   d. thigmotropic

8. Abscission occurs during _____ .
   a. seed germination   c. senescence
   b. flowering   d. dormancy

9. Senescence involves a decrease in _____ in leaves and fruits and an increase in _____ at abscission zones.
   a. IAA; ethylene   c. Pfr; gibberellin
   b. ethylene; IAA   d. gibberellin; abscisic acid

10. Match the plant reproduction and development terms.
   ____ vernalization   a. water moves into seeds
   ____ senescence   b. unequal growth following
   ____ imbibition   contact with solid objects
   ____ thigmotropism   c. lateral bud formation inhibited
   ____ apical   d. low-temperature stimulation
        dominance   of the flowering process
                             e. all processes leading to death
                                of plant or plant part

Critical Thinking

1. Given what you know about the growth of plants (Chapter 29), would you expect hormones to influence primary growth only? What about secondary growth in, say, a redwood tree?

2. Plant growth depends on photosynthesis, which depends on inputs of light energy from the sun. How, then, can seedlings that were germinated in a dark room grow taller than different seedlings that germinated in the sun?

3. *Solar tracking* refers to the observation that many plants are able to maintain the flat blades of their leaves at right angles to the sun throughout the day. Figure 32.23 gives an example. This tropic response maximizes the harvesting of the sun's rays by leaves. Suggest the name of one type of molecule that might be involved in the response.

4. Belgian scientists isolated a mutated gene in wall cress (*Arabidopsis thaliana*) that produces excess amounts of auxin. Predict what some of the resulting phenotypic traits might be.

5. Remember Section 28.5? All flowers are variations on a basic pattern of growth and development. Two groups led by Elliot Meyerowitze and Detlef Weigel recently identified the genes responsible for that plan in *Arabidopsis*. The master gene (*leafy*) activates other genes that contribute to the formation of sepals (gene *A*), petals (gene *B*), and reproductive structures (gene *C*). Discovering these genes and their interactions has been called the botanical equivalent of isolating master genes of *Drosophila* development. Speculate on what kind of internal and external signals switch on the leafy gene in the first place.

6. Cattle typically are given somatotropin, an animal hormone that makes them grow bigger (the added weight means greater profits). There is a major concern that such hormones may have unforeseen side effects on beef-eating humans. Would you think plant hormones applied to crop plants can affect humans also? Why or why not?

Selected Key Terms

abscisic acid *32.2*	ethylene *32.2*	photoperiodism *32.4*
abscission *32.5*	flavoprotein *32.3*	phototropism *32.3*
auxin *32.2*	germination *32.1*	phytochrome *32.4*
biological clock *32.4*	gibberellin *CI*	senescence *32.5*
coleoptile *32.2*	gravitropism *32.3*	short-day
cytokinin *32.2*	growth *32.2*	plant *32.4*
day-neutral	herbicide *32.2*	statolith *32.3*
plant *32.4*	hormone *CI*	thigmotropism *32.3*
development *32.2*	imbibition *32.1*	toxin *32.6*
dormancy *32.5*	long-day plant *32.4*	vernalization *32.5*

Readings

Meyerowitz, E. M. November 1994. "Genetics of Flower Development." *Scientific American* 271: 56–65.

Raven, P. , R. Evert, and S. Eichhorn. 1999. *Biology of Plants*. Sixth edition. New York: Freeman/Worth.

Rost, T., M. Barbour, C. Stocking, and T. Murphy. 1998. *Plant Biology*. Belmont, California: Wadsworth. Paperback.

Salisbury, F., and C. Ross. 1992. *Plant Physiology*. Fourth edition. Belmont, California: Wadsworth.

On-Line readings at Student Guide for InfoTrac:
www.brookscole.com/biology

APPENDIX I.   CLASSIFICATION SYSTEM

This revised classification scheme is a composite of several that microbiologists, botanists, and zoologists use. The major groupings are agreed upon, more or less. However, there is not always agreement on what to name a particular grouping or where it might fit within the overall hierarchy. There are several reasons why full consensus is not possible at this time.

First, the fossil record varies in its completeness and quality (Section 19.1). As one outcome, the phylogenetic relationship of one group to other groups is sometimes open to interpretation. Today, comparative studies at the molecular level are firming up the picture, but the work is still under way.

Second, ever since the time of Linnaeus, systems of classification have been based on the perceived morphological similarities and differences among organisms. Although some original interpretations are now open to question, we are so used to thinking about organisms in certain ways that reclassification often proceeds slowly.

A few examples: Traditionally, birds and reptiles were grouped in separate classes (Reptilia and Aves); yet there are many compelling arguments for grouping the lizards and snakes in one class and the crocodilians, dinosaurs, and birds in a separate class. Some biologists have favored a six-kingdom system of classification (archaebacteria, eubacteria, protistans, plants, fungi, and animals). Many others now favor a three-domain classification system. The archaebacteria, eubacteria, and eukaryotes (alternatively, the archaea, bacteria, and eukarya) are its major groupings.

Third, researchers in microbiology, mycology, botany, zoology, and other fields of inquiry inherited a wealth of literature, based on classification systems that have been developed over time in each field of inquiry. Many simply do not wish to give up established terminology that offers access to the past.

For example, botanists and microbiologists often use *division*, and zoologists *phylum*, for taxa that actually are equivalent in hierarchies of classification. As another example, opinions are quite polarized with respect to kingdom Protista, certain members of which could easily be grouped with plants, or fungi, or animals. Indeed, the term "protozoan" is a holdover from an earlier scheme in which some single-celled organisms were ranked as simple animals.

Given the problems, why do systematists work so hard to construct hypotheses regarding the history of life? They do so because classification systems that accurately reflect evolutionary relationships are more useful, with far more predictive power for comparative studies. Such systems also serve as a good framework for studying the living world, which could otherwise be an overwhelming body of knowledge. Importantly, classifications enhance the retrieval of information that relates to living organisms.

Bear in mind, *we include this appendix on classification for your reference purposes only*. Besides being open to revision, it is by no means complete. Names in "quotes" are polyphyletic or paraphyletic groups undergoing revision. The groupings of certain animal phyla reflect common ancestry (Nematoda through Arthropoda, and Nemertea through Annelida), as emerging molecular data suggest. Also, the most recently discovered species, as from the mid-ocean province, are not listed. Many existing and extinct species of the more obscure phyla are not represented. Our strategy is to focus primarily on the organisms mentioned in the text.

PROKARYOTES AND EUKARYOTES COMPARED

As a general frame of reference, note that almost all eubacteria and archaebacteria are microscopic in size. Their DNA is concentrated in a nucleoid (a region of cytoplasm), not in a membrane-bound nucleus. All are single cells or simple associations of cells. They reproduce by prokaryotic fission or budding; they transfer genes by bacterial conjugation.

Table A lists representative types of autotrophic and heterotrophic prokaryotes. The authoritative reference, *Bergey's Manual of Systematic Bacteriology,* has called this a time of taxonomic transition. It references groups mainly by numerical taxonomy (Section 21.3) rather than by phylogeny. Our classification system does reflect evidence of evolutionary relationships for at least some bacterial groups.

The first life forms were prokaryotic. Similarities between Eubacteria and Archaebacteria have more ancient origins relative to the traits of eukaryotes.

Unlike the prokaryotes, all eukaryotic cells start out life with a DNA-enclosing nucleus and other membrane-bound organelles. Their chromosomes have many histones and other proteins attached. They include spectacularly diverse single-celled and multicelled species, which can reproduce by way of meiosis, mitosis, or both.

BACTERIA — EUBACTERIA

ARCHAEA — ARCHAEBACTERIA

EUKARYA — PROTISTA   FUNGI   PLANTAE   ANIMALIA

DOMAIN OF EUBACTERIA (BACTERIA)

KINGDOM EUBACTERIA   Gram-negative and Gram-positive prokaryotic cells. Peptidoglycan in cell wall. Collectively, great metabolic diversity; photosynthetic autotrophs, chemosynthetic autotrophs, and heterotrophs.

PHYLUM FIRMICUTES   Typically Gram-positive, thick wall. Heterotrophs. *Bacillus, Staphylococcus, Streptococcus, Clostridium, Actinomycetes.*

PHYLUM GRACILICUTES   Typically Gram-negative, thin wall. Autotrophs (photosynthetic and chemosynthetic) and heterotrophs. *Anabaena* and other cyanobacteria. *Escherichia, Pseudomonas, Neisseria, Myxococcus.*

PHYLUM TENERICUTES   Gram-negative, wall absent. Heterotrophs (saprobes, pathogens). *Mycoplasma.*

DOMAIN OF ARCHAEBACTERIA (ARCHAEA)

KINGDOM ARCHAEBACTERIA   Methanogens, extreme halophiles, extreme thermophiles. Evolutionarily closer to eukaryotic cells than to eubacteria. All strict anaerobes living in habitats as harsh as those that probably prevailed on the early Earth. Compared with other prokaryotic cells, all archaebacteria have a distinctive cell wall and unique membrane lipids, ribosomes, and RNA sequences. *Methanobacterium, Halobacterium, Sulfolobus.*

Table A   Representative Eubacteria and Archaebacteria Grouped on the Basis of Numerical Taxonomy

Some Major Groups	Main Habitats	Characteristics	Representatives
EUBACTERIA			
Photoautotrophs:			
Cyanobacteria, green sulfur bacteria, and purple sulfur bacteria	Mostly lakes, ponds; some marine, terrestrial habitats	Photosynthetic; use sunlight energy, carbon dioxide; cyanobacteria use oxygen-producing noncyclic pathway; some also use cyclic route	*Anabaena, Nostoc, Rhodopseudomonas, Chloroflexus*
Photoheterotrophs:			
Purple nonsulfur and green nonsulfur bacteria	Anaerobic, organically rich muddy soils, and sediments of aquatic habitats	Use sunlight energy; organic compounds as electron donors; some purple nonsulfur may also grow chemotrophically	*Rhodospirillum, Chlorobium*
Chemoautotrophs:			
Nitrifying, sulfur-oxidizing, and iron-oxidizing bacteria	Soil; freshwater, marine habitats	Use carbon dioxide, inorganic compounds as electron donors; influence crop yields, cycling of nutrients in ecosystems	*Nitrosomonas, Nitrobacter, Thiobacillus*
Chemoheterotrophs:			
Spirochetes	Aquatic habitats; parasites of animals	Helically coiled, motile; free-living and parasitic species; some major pathogens	*Spirochaeta, Treponema*
Gram-negative aerobic rods and cocci	Soil, aquatic habitats; parasites of animals, plants	Some major pathogens; some fix nitrogen (e.g., *Rhizobium*)	*Pseudomonas, Neisseria, Rhizobium, Agrobacterium*
Gram-negative facultative anaerobic rods	Soil, plants, animal gut	Many major pathogens; one bioluminescent (*Photobacterium*)	*Salmonella, Escherichia, Proteus, Photobacterium*
Rickettsias and chlamydias	Host cells of animals	Intracellular parasites; many pathogens	*Rickettsia, Chlamydia*
Myxobacteria	Decaying organic material; bark of living trees	Gliding, rod-shaped; aggregation and collective migration of cells	*Myxococcus*
Gram-positive cocci	Soil; skin and mucous membranes of animals	Some major pathogens	*Staphylococcus, Streptococcus*
Endospore-forming rods and cocci	Soil; animal gut	Some major pathogens	*Bacillus, Clostridium*
Gram-positive nonsporulating rods	Fermenting plant, animal material; gut, vaginal tract	Some important in dairy industry, others major contaminators of milk, cheese	*Lactobacillus, Listeria*
Actinomycetes	Soil; some aquatic habitats	Include anaerobes and strict aerobes; major producers of antibiotics	*Actinomyces, Streptomyces*
ARCHAEBACTERIA (ARCHAEA)			
Methanogens	Anaerobic sediments of lakes, swamps; animal gut	Chemosynthetic; methane producers; used in sewage treatment facilities	*Methanobacterium*
Extreme halophiles	Brines (extremely salty water)	Heterotrophic; also, unique photosynthetic pigments (bacteriorhodopsin) form in some	*Halobacterium*
Extreme thermophiles	Acidic soil, hot springs, hydrothermal vents	Heterotrophic or chemosynthetic; use inorganic substances as electron donors	*Sulfolobus, Thermoplasma*

DOMAIN OF EUKARYOTES (EUKARYA)

KINGDOM "PROTISTA"   Diverse single-celled, colonial, and multicelled eukaryotic species. Existing types are unlike prokaryotes and most like the earliest forms of eukaryotes. Autotrophs, heterotrophs, or both (Table 22.1). Reproduce sexually and asexually (by meiosis, mitosis, or both). Not a monophyletic group. The kingdom may soon be split into multiple kingdoms, and some of its groups are already being reclassified as plants or fungi.

PHYLUM "MASTIGOPHORA" Flagellated protozoans. Free-living heterotrophs; many are internal parasites. They have one to several flagella. At present, a non-monophyletic grouping of ancient lineages, including the diplomonads, parabasalids, and kinetoplastids. *Trypanosoma, Trichomonas, Giardia.*

PHYLUM EUGLENOPHYTA Euglenoids. Mostly heterotrophs, some photoautotrophs, some both depending on conditions. Most with one short, one long flagellum. Pigmented (red, green) or colorless. Related to kinetoplastids. *Euglena.*

PHYLUM SARCODINA Amoeboid protozoans. Heterotrophs, free-living or endosymbionts, some pathogens. Soft-bodied, with or without shell, pseudopods. Rhizopods (naked amoebas, foraminiferans), actinopods (radiolarians, heliozoans). *Amoeba.*

ALVEOLATES

PHYLUM CILIOPHORA Ciliated protozoans. Heterotrophs, predators or symbionts, some parasitic. All have cilia. Free-living, sessile, or motile. *Paramecium, Didinium,* hypotrichs.

PHYLUM APICOMPLEXA Heterotrophs, sporozoite-forming parasites. Complex structures at head end. Most familiar types known as sporozoans. *Cryptosporidium, Plasmodium, Toxoplasma.*

PHYLUM PYRRHOPHYTA. Dinoflagellates. Photosynthetic, mostly, but some heterotrophs. *Pfiesteria, Gymnodinium breve.*

STRAMENOPILES

PHYLUM OOMYCOTA. Water molds. Heterotrophs. Decomposers, some parasites. *Saprolegnia, Phytophthora, Plasmopara.*

PHYLUM CHRYSOPHYTA. Golden algae, yellow-green algae, diatoms. Photosynthetic. Some flagellated. *Mischococcus, Synura, Vaucheria.*

PHYLUM PHAEOPHYTA. Brown algae. Photosynthetic, nearly all endemic to temperate or marine waters. *Macrocystis, Fucus, Sargassum, Ectocarpus, Postelsia.*

GROUPS CLOSELY RELATED TO PLANTS

PHYLUM CHLOROPHYTA Green algae. Mostly photosynthetic, some parasitic. Most freshwater, some marine or terrestrial. *Chlamydomonas, Spirogyra, Ulva, Volvox, Codium, Halimeda.*

PHYLUM RHODOPHYTA Red algae. Mostly photosynthetic, some parasitic. Nearly all marine, some in freshwater habitats. *Porphyra, Bonnemaisonia, Euchema.*

PHYLUM CHAROPHYTA Stoneworts. *Chara.*

GROUPS CLOSELY RELATED TO FUNGI

PHYLUM CHYTRIDIOMYCOTA Chytrids. Heterotrophs; saprobic decomposers or parasites. *Chytridium.*

GROUPS OF SLIME MOLDS

PHYLUM ACRASIOMYCOTA Cellular slime molds. Heterotrophs with free-living, phagocytic amoeboid cells and spore-bearing stages. *Dictyostelium.*

PHYLUM MYXOMYCOTA Plasmodial slime molds. Heterotrophs with free-living, phagocytic amoeboid cells and spore-bearing stages. Aggregate into streaming mass of cells that discard their plasma membrane. *Physarum.*

KINGDOM PLANTAE   Multicelled eukaryotes. Nearly all photosynthetic autotrophs with chlorophylls *a* and *b*. Some parasitic. Nonvascular and vascular species, generally with well-developed root and shoot systems. Nearly all adapted to survive dry conditions on land; a few in aquatic habitats. Sexual reproduction predominant with spore-forming chambers and embryos in life cycle; also asexual reproduction by vegetative propagation and other mechanisms.

PHYLUM "BRYOPHYTA" Bryophytes; mosses, liverworts, hornworts. Not a monophyletic group. Seedless, nonvascular, haploid dominance. *Marchantia, Polytrichum, Sphagnum.*

VASCULAR PLANTS

PHYLUM "RHYNIOPHYTA" Earliest known vascular plants; muddy habitats. A polyphyletic group, some are primitive lycophytes. Extinct. *Cooksonia, Rhynia.*

PHYLUM LYCOPHYTA Lycophytes, club mosses. Seedless, vascular. Small leaves, branching rhizomes, vascularized roots and stems. *Lepidodendron* (extinct), *Lycopodium, Selaginella.*

PHYLUM SPHENOPHYTA Horsetails. Seedless, vascular, whorled leaves. Some stems photosynthetic, others nonphotosynthetic, spore-producing. *Calamites* (extinct), *Equisetum.*

PHYLUM CYCADOPHYTA Cycads. Gymnosperm group (vascular, bear "naked" seeds). Tropical, subtropical. Palm-shaped leaves, simple cones on male and female plants. *Zamia.*

PHYLUM PTEROPHYTA. Ferns. Large leaves, usually with sori. Largest group of seedless vascular plants (12,000 species), mainly tropical, temperate habitats. *Pteris, Trichomanes, Cyathea* (tree ferns), *Polystichum.*

PHYLUM PSILOPHYTA Whisk ferns. Seedless, vascular. No obvious roots, leaves on sporophyte, very reduced. *Psilotum.*

PHYLUM "PROGYMNOSPERMOPHYTA" The progymnosperms. Ancestral to early seed-bearing plants; extinct. *Archaeopteris.*

SEED-BEARING PLANTS (A subgroup of vascular plants)

PHYLUM "PTERIDOSPERMOPHYTA" Seed ferns. Fernlike gymnosperms; extinct. *Medullosa.*

PHYLUM CYCADOPHYTA Cycads. Group of gymnosperms (vascular, bear "naked" seeds). Tropical, subtropical. Compound leaves, simple cones on male and female plants. Plants usually palm-like. *Zamia, Cycas.*

PHYLUM GINKGOPHYTA Ginkgo (maidenhair tree). Type of gymnosperm. Seeds with fleshy outer layer. *Ginkgo.*

PHYLUM GNETOPHYTA Gnetophytes. Only gymnosperms with vessels in xylem and double fertilization (but endosperm does not form). *Ephedra, Welwitschia, Gnetum.*

PHYLUM CONIFEROPHYTA Conifers. Most common and familiar gymnosperms. Generally cone-bearing species with needle-like or scale-like leaves.

Family Pinaceae. Pines (*Pinus*), firs (*Abies*), spruces (*Picea*), hemlock (*Tsuga*), larches (*Larix*), true cedars (*Cedrus*).

Family Cupressaceae. Junipers (*Juniperus*), Cypresses (*Cupressus*), Bald cypress (*Taxodium*), redwood (*Sequoia*), bigtree (*Sequoiadendron*), dawn redwood (*Metasequoia*).

Family Taxaceae. Yews. *Taxus.*

PHYLUM ANTHOPHYTA Angiosperms (the flowering plants). Largest, most diverse group of vascular seed-bearing plants. Only organisms that produce flowers, fruits. Some families from several representative orders are listed:

BASAL FAMILIES

Family Amborellaceae. *Amborella.*
Family Nymphaeaceae. Water lilies.
Family Illiciaceae. Star anise.

MAGNOLIIDS

Family Magnoliaceae. Magnolias.
Family Lauraceae. Cinnamon, sassafras, avocados.
Family Piperaceae. Black pepper, white pepper.

EUDICOTS

Family Papaveraceae. Poppies.
Family Cactaceae. Cacti.
Family Euphorbiaceae. Spurges, poinsettia.
Family Salicaceae. Willows, poplars.
Family Fabaceae. Peas, beans, lupines, mesquite.
Family Rosaceae. Roses, apples, almonds, strawberries.
Family Moraceae. Figs, mulberries.
Family Cucurbitaceae. Gourds, melons, cucumbers, squashes.
Family Fagaceae. Oaks, chestnuts, beeches.
Family Brassicaceae. Mustards, cabbages, radishes.
Family Malvaceae. Mallows, okra, cotton, hibiscus, cocoa.
Family Sapindaceae. Soapberry, litchi, maples.
Family Ericaceae. Heaths, blueberries, azaleas.
Family Rubiaceae. Coffee.
Family Lamiaceae. Mints.
Family Solanaceae. Potatoes, eggplant, petunias.
Family Apiaceae. Parsleys, carrots, poison hemlock.
Family Asteraceae. Composites. Chrysanthemums, sunflowers, lettuces, dandelions.

MONOCOTS

Family Araceae. Anthuriums, calla lily, philodendrons.
Family Liliaceae. Lilies, tulips.
Family Alliaceae. Onions, garlic.
Family Iridaceae. Irises, gladioli, crocuses.
Family Orchidaceae. Orchids.
Family Arecaceae. Date palms, coconut palms.
Family Bromeliaceae. Bromeliads, pineapples, Spanish moss.
Family Cyperaceae. Sedges.
Family Poaceae. Grasses, bamboos, corn, wheat, sugarcane.
Family Zingiberaceae. Gingers.

FUNGI

Nearly all multicelled eukaryotic species. Heterotrophs, mostly saprobic decomposers, some parasites. Nutrition based upon extracellular digestion of organic matter and absorption of nutrients by individual cells. Multicelled species form absorptive mycelia within substrates and structures that produce asexual spores (and sometimes sexual spores).

PHYLUM ZYGOMYCOTA   Zygomycetes. Producers of zygospores (zygotes inside thick wall) by way of sexual reproduction. Bread molds, related forms. *Rhizopus, Philobolus.*

PHYLUM ASCOMYCOTA   Ascomycetes. Sac fungi. Sac-shaped cells form sexual spores (ascospores). Most yeasts and molds, morels, truffles. *Saccharomyces, Morchella, Neurospora, Sarcoscypha, Claviceps, Ophiostoma, Candida, Aspergillus, Penicillium.*

PHYLUM BASIDIOMYCOTA   Basidiomycetes. Club fungi. Most diverse group. Produce basidiospores inside club-shaped structures. Mushrooms, shelf fungi, stinkhorns. *Agaricus, Amanita, Craterellus, Gymnophilus, Puccinia, Ustilago.*

"IMPERFECT FUNGI"   Sexual spores absent or undetected. The group has no formal taxonomic status. If better understood, a given species might be grouped with sac fungi or club fungi. *Arthobotrys, Histoplasma, Microsporum, Verticillium.*

"LICHENS"   Mutualistic interactions between fungal species and a cyanobacterium, green alga, or both. *Lobaria, Usnea, Cladonia.*

KINGDOM ANIMALIA

Multicelled eukaryotes, nearly all with tissues, organs, and organ systems; show motility during at least part of their life cycle; embryos develop through a series of stages. Diverse heterotrophs, predators (herbivores, carnivores, omnivores), parasites, detritivores. Reproduce sexually and, in many species, asexually as well.

PHYLUM PORIFERA   Sponges. No symmetry, tissues. *Euplectella.*

PHYLUM PLACOZOA   Marine. Simplest known animal. Two cell layers, no mouth, no organs. *Trichoplax.*

PHYLUM CNIDARIA   Radial symmetry, tissues, nematocysts.
   Class Hydrozoa. Hydrozoans. *Hydra, Obelia, Physalia, Prya.*
   Class Scyphozoa. Jellyfishes. *Aurelia.*
   Class Anthozoa. Sea anemones, corals. *Telesto.*

PHYLUM MESOZOA   Ciliated, wormlike parasites, about the same level of complexity as *Trichoplax.*

PHYLUM PLATYHELMINTHES   Flatworms. Bilateral, cephalized; simplest animals with organ systems. Saclike gut.
   Class Turbellaria. Triclads (planarians), polyclads. *Dugesia.*
   Class Trematoda. Flukes. *Clonorchis, Schistosoma.*
   Class Cestoda. Tapeworms. *Diphyllobothrium, Taenia.*

PHYLUM ROTIFERA   Rotifers. *Asplancha, Philodina.*

PHYLUM NEMERTEA   Ribbon worms. *Tubulanus.*

PHYLUM MOLLUSCA   Mollusks.
   Class Polyplacophora. Chitons. *Cryptochiton, Tonicella.*
   Class Gastropoda. Snails (periwinkles, whelks, limpets, abalones, cowries, conches, nudibranchs, tree snails, garden snails), sea slugs, land slugs. *Aplysia, Ariolimax, Cypraea, Haliotis, Helix, Liguus, Limax, Littorina, Patella.*
   Class Bivalvia. Clams, mussels, scallops, cockles, oysters, shipworms. *Ensis, Chlamys, Mytelus, Patinopectin.*
   Class Cephalopoda. Squids, octopuses, cuttlefish, nautiluses. *Dosidiscus, Loligo, Nautilus, Octopus, Sepia.*

PHYLUM BRYOZOA   Bryozoans (moss animals).

PHYLUM BRACHIOPODA   Lampshells.

PHYLUM ANNELIDA   Segmented worms.
   Class Polychaeta. Mostly marine worms. *Eunice, Neanthes.*
   Class Oligochaeta. Mostly freshwater and terrestrial worms, many marine. *Lumbricus* (earthworms), *Tubifex.*
   Class Hirudinea. Leeches. *Hirudo, Placobdella.*

PHYLUM NEMATODA   Roundworms. *Ascaris, Caenorhabditis elegans, Necator* (hookworms), *Trichinella.*

PHYLUM TARDIGRADA   Water bears.

PHYLUM ONYCHOPHORA   Onychophorans. *Peripatus.*

PHYLUM ARTHROPODA.
   Subphylum Trilobita. Trilobites; extinct.
   Subphylum Chelicerata. Chelicerates. Horseshoe crabs, spiders, scorpions, ticks, mites.
   Subphylum Crustacea. Shrimps, crayfishes, lobsters, crabs, barnacles, copepods, isopods (sowbugs).
   Subphylum Uniramia.
     Superclass Myriapoda. Centipedes, millipedes.
     Superclass Insecta.
        Order Ephemeroptera. Mayflies.
        Order Odonata. Dragonflies, damselflies.
        Order Orthoptera. Grasshoppers, crickets, katydids.
        Order Dermaptera. Earwigs.

Order Blattodea. Cockroaches.
Order Mantodea. Mantids.
Order Isoptera. Termites.
Order Mallophaga. Biting lice.
Order Anoplura. Sucking lice.
Order Hemiptera. Cicadas, aphids, leafhoppers, spittlebugs, bugs.
Order Coleoptera. Beetles.
Order Diptera. Flies.
Order Mecoptera. Scorpion flies. *Harpobittacus.*
Order Siphonaptera. Fleas.
Order Lepidoptera. Butterflies, moths.
Order Hymenoptera. Wasps, bees, ants.
Order Neuroptera. Lacewings, antlions.

PHYLUM ECHINODERMATA. Echinoderms.

Class Asteroidea. Sea stars. *Asterias.*
Class Ophiuroidea. Brittle stars.
Class Echinoidea. Sea urchins, heart urchins, sand dollars.
Class Holothuroidea. Sea cucumbers.
Class Crinoidea. Feather stars, sea lilies.
Class Concentricycloidea. Sea daisies.

PHYLUM HEMICHORDATA. Acorn worms.

PHYLUM CHORDATA. Chordates.

Subphylum Urochordata. Tunicates, related forms.

Subphylum Cephalochordata. Lancelets.

CRANIATES

Superclass "Agnatha." Jawless fishes, including ostracoderms (extinct).

Class Myxini. Hagfishes.

Class Cephalaspidomorphi. Lampreys.

Subphylum Vertebrata. Jawed vertebrates.

Class "Placodermi." Jawed, heavily armored fishes; extinct.

Class Chondrichthyes. Cartilaginous fishes (sharks, rays, skates, chimaeras).

Class "Osteichthyes." Bony fishes. Not monophyletic.
Subclass Dipnoi. Lungfishes.
Subclass Crossopterygii. Coelacanths, related forms.
Subclass Actinopterygii. Ray-finned fishes.
Order Acipenseriformes. Sturgeons, paddlefishes.
Order Salmoniformes. Salmon, trout.
Order Atheriniformes. Killifishes, guppies.
Order Gasterosteiformes. Seahorses.
Order Perciformes. Perches, wrasses, barracudas, tunas, freshwater bass, mackerels.
Order Lophiiformes. Angler fishes.

TETRAPODS (A subgroup of craniates)

Class Amphibia. Amphibians.
Order Caudata. Salamanders.
Order Anura. Frogs, toads.
Order Apoda. Apodans (caecilians).

AMNIOTES (A subgroup of tetrapods)

Class "Reptilia." Skin with scales, embryo protected and nutritionally supported by extraembryonic membranes.
Subclass Anapsida. Turtles, tortoises.
Subclass Lepidosaura. *Sphenodon,* lizards, snakes.
Subclass Archosaura. Dinosaurs (extinct), crocodiles, alligators.

Class Aves. Birds. In some of the more recent classification systems, dinosaurs, crocodilians, and birds are grouped in the same category, the archosaurs.
Order Struthioniformes. Ostriches.
Order Sphenisciformes. Penguins.
Order Procellariiformes. Albatrosses, petrels.
Order Ciconiiformes. Herons, bitterns, storks, flamingoes.
Order Anseriformes. Swans, geese, ducks.
Order Falconiformes. Eagles, hawks, vultures, falcons.
Order Galliformes. Ptarmigan, turkeys, domestic fowl.
Order Columbiformes. Pigeons, doves.
Order Strigiformes. Owls.
Order Apodiformes. Swifts, hummingbirds.
Order Passeriformes. Sparrows, jays, finches, crows, robins, starlings, wrens.
Order Piciformes. Woodpeckers, toucans.
Order Psittaciformes. Parrots, cockatoos, macaws.

Class Mammalia. Skin with hair; young nourished by milk-secreting glands of adult.

Subclass Prototheria. Egg-laying mammals (monotremes; duckbilled platypus, spiny anteaters).

Subclass Metatheria. Pouched mammals or marsupials (opossums, kangaroos, wombats, Tasmanian devil).

Subclass Eutheria. Placental mammals.
Order Edentata. Anteaters, tree sloths, armadillos.
Order Insectivora. Tree shrews, moles, hedgehogs.
Order Chiroptera. Bats.
Order Scandentia. Insectivorous tree shrews.
Order Primates.
Suborder Strepsirhini (prosimians). Lemurs, lorises.
Suborder Haplorhini (tarsioids and anthropoids).
Infraorder Tarsiiformes. Tarsiers.
Infraorder Platyrrhini (New World monkeys).
Family Cebidae. Spider monkeys, howler monkeys, capuchin.
Infraorder Catarrhini (Old World monkeys and hominoids).
Superfamily Cercopithecoidea. Baboons, macaques, langurs.
Superfamily Hominoidea. Apes and humans.
Family Hylobatidae. Gibbon.
Family "Pongidae." Chimpanzees, gorillas, orangutans.
Family Hominidae. Existing and extinct human species (*Homo*) and humanlike species, including the australopiths.
Order Lagomorpha. Rabbits, hares, pikas.
Order Rodentia. Most gnawing animals (squirrels, rats, mice, guinea pigs, porcupines, beavers, etc.).
Order Carnivora. Carnivores.
Suborder Feloidea. Cats, mongooses, hyenas.
Suborder Canoidea. Dogs, weasels, skunks, otters, raccoons, pandas, bears.
Order Pinnipedia. Seals, walruses, sea lions.
Order Proboscidea. Elephants; mammoths (extinct).
Order Sirenia. Sea cows (manatees, dugongs).
Order Perissodactyla. Odd-toed ungulates (horses, tapirs, rhinos).
Order Tubulidentata. African aardvarks.
Order Artiodactyla. Even-toed ungulates (camels, deer, bison, sheep, goats, antelopes, giraffes, etc.).
Order Cetacea. Whales, porpoises.

Metric-English Conversions

Length

English		Metric
inch	=	2.54 centimeters
foot	=	0.30 meter
yard	=	0.91 meter
mile (5,280 feet)	=	1.61 kilometer

To convert	multiply by	to obtain
inches	2.54	centimeters
feet	30.00	centimeters
centimeters	0.39	inches
millimeters	0.039	inches

Weight

English		Metric
grain	=	64.80 milligrams
ounce	=	28.35 grams
pound	=	453.60 grams
ton (short) (2,000 pounds)	=	0.91 metric ton

To convert	multiply by	to obtain
ounces	28.3	grams
pounds	453.6	grams
pounds	0.45	kilograms
grams	0.035	ounces
kilograms	2.2	pounds

Volume

English		Metric
cubic inch	=	16.39 cubic centimeters
cubic foot	=	0.03 cubic meter
cubic yard	=	0.765 cubic meters
ounce	=	0.03 liter
pint	=	0.47 liter
quart	=	0.95 liter
gallon	=	3.79 liters

To convert	multiply by	to obtain
fluid ounces	30.00	milliliters
quart	0.95	liters
milliliters	0.03	fluid ounces
liters	1.06	quarts

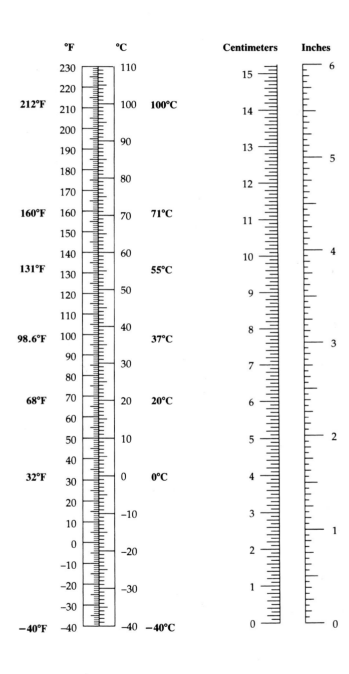

APPENDIX III. ANSWERS TO SELF-QUIZZES

Italicized numbers refer to relevant page numbers

CHAPTER 28		
1.	a	*490*
2.	c	*491*
3.	e	*491*
4.	d	*498*
5.	a	*489*
	c	*497*
	d	*491, 494*
	b	*489, 498*
	e	*494*

CHAPTER 29		
1.	a	*507*
2.	d	*507, 516, 518*
3.	a	*508*
4.	b	*508*
5.	c	*508*
6.	d	*510*
7.	c	*513*
8.	d	*519*
9.	b	*507*
	d	*507*
	e	*508*
	c	*509, 518*
	f	*514–515*
	a	*516–517*

CHAPTER 30		
1.	e	*524, 524t*
2.	c	*526*
3.	b	*527*
4.	a	*527*
5.	hydrogen bonds	*528–529*
6.	c	*528*
7.	d	*528–529*
8.	a	*530*
9.	d	*530*
10.	b	*532*
11.	c	*530–531*
	g	*524*
	e	*532*
	b	*526*
	d	*528–529*
	a	*528*
	f	*532*

CHAPTER 31		
1.	pollinators	*536*
2.	a	*538*
3.	d	*538, 540*
4.	b	*538i, 539*
5.	c	*541–543*
6.	b	*540*
7.	b	*540*
8.	a	*543*
9.	b	*542*
10.	d	*543*
11.	c	*546t*
12.	a	*541*
	c	*540*
	d	*540, 540i*
	e	*546*
	b	*536*

CHAPTER 32		
1.	c	*552*
2.	c	*554t, 455*
3.	e	*550–553*
4.	d	*554*
5.	d	*557*
6.	a	*558*
7.	c	*558*
8.	c	*560*
9.	a	*560*
10.	d	*561*
	e	*560*
	a	*552*
	b	*557*
	c	*555*

GLOSSARY

ABC model Idea that products of three groups of master genes control floral development.

abscisic acid Plant hormone; induces stomatal closure (conserves water); bud, seed dormancy.

abscission (ab-SIH-zhun) Dropping of leaves, flowers, fruits, or other parts from a plant.

active transport Pumping of a specific solute across a cell membrane against its concentration gradient, through a transport protein's interior. Requires an energy boost, typically from ATP.

adaptation [L. *adaptare*, to fit] Long-term, heritable aspect of form, function, behavior, or development that improves odds for surviving and reproducing; an outcome of microevolution (e.g., natural selection that enhances fit between individual and prevailing conditions). Of sensory neurons, a decline or end of action potentials even when a stimulus is maintained at constant strength.

anatomy Study of body form, from molecular to organ-system levels of organization.

annual Plant that lasts one growing season.

auxin (AWK-sin) Plant hormone; induces stem lengthening and responses to gravity and light.

bark All tissues external to vascular cambium.

biennial (bi-EN-yul) Flowering plant that completes its life cycle in two growing seasons.

biological clock Internal time-measuring mechanism; adjusts activities seasonally, daily, or both in response to environmental cues.

bud Undeveloped shoot, mainly meristematic tissue. Small, protective scales often cover it.

CAM plant Type of plant that conserves water by opening stomata only at night, when it fixes carbon dioxide by means of a C4 pathway.

carnivorous plant Plant that traps, digests, and feeds on insects and other small animals.

carpel (KAR-pul) Female reproductive part of a flower. The ovary, stigma, and often a style consist of one or more carpels.

Casparian strip Narrow, waxy, impermeable band between walls of abutting cells making up root endodermis and exodermis.

circadian rhythm (ser-KAYD-ee-un) [L. *circa*, about, + *dies*, day] Biological activity repeated in cycles, each about twenty-four hours long, independently of environmental change.

coevolution Joint evolution of two closely interacting species by changes in the selection pressures operating between the two.

cohesion Capacity to resist rupturing when placed under tension (stretched).

cohesion theory of water transport Theory that collective cohesive strength of hydrogen bonds pulls up water molecules through xylem in response to transpiration from leaves.

coleoptile Thin sheath that protects primary shoot as growth pushes it up through soil.

collenchyma (coll-ENG-kih-mah) A simple plant tissue that imparts flexible support during primary growth, as in lengthening stems.

companion cell Specialized parenchyma cell; helps load conducting cells of phloem.

compartmentalization Plant responses to an attack, including secretion of resins and toxins.

cork Periderm component; insulates, protects, and waterproofs woody stems and roots.

cork cambium Lateral meristem that replaces epidermis with cork on woody plant parts.

cortex A rindlike layer. In vascular plants, a ground tissue; supports parts and stores food.

cotyledon (KOT-uhl-EE-dun) Seed leaf. One develops in monocot seeds and has digestive roles; two develop in dicot seeds and store food for germination and early growth.

cuticle (KEW-tih-kull) Body cover. Of plants, a transparent covering of waxes and cutin on outer epidermal cell walls. Of annelids, a thin, flexible coat. Of arthropods, a lightweight exoskeleton hardened with protein and chitin.

cutin Insoluble lipid polymer that functions as the ground substance for plant cuticles.

cytokinin (SIGH-toe-KYE-nin) Type of plant hormone; stimulates cell division, promotes leaf expansion, and retards leaf aging.

day-neutral plant Plant that flowers when mature, independently of day or night length.

dermal tissue system Tissues that cover and protect all exposed surfaces of a plant.

development The series of genetically guided embryonic and post-embryonic stages by which morphologically distinct, specialized body parts emerge in a new multicelled individual.

diffusion Net movement of like molecules or ions down their concentration gradient.

division of labor A splitting up of tasks among cells, tissues, organs, organ systems that collectively contribute to the survival and reproduction of a multicelled organism.

dormancy [L. *dormire*, to sleep] A predictable time of metabolic inactivity for many spores, cysts, seeds, perennials, and some animals.

double fertilization Of flowering plants only, fusion of a sperm and egg nucleus, plus fusion of another sperm nucleus with nuclei of a cell that gives rise to a nutritive tissue (endosperm).

effector Muscle (or gland); helps bring about movement (or chemical change) in response to neural or endocrine signals.

endodermis Cell layer around root vascular cylinder; influences water and solute uptake.

endosperm (EN-doe-sperm) Nutritive tissue inside the seed of a flowering plant.

epidermis Outermost tissue layer of plants and all animals above sponge level of organization.

erosion Movement of land under the force of wind, running water, and ice.

ethylene (ETH-il-een) Plant hormone; promotes fruit ripening and leaf, flower, fruit abscission.

eudicot True dicot; one of three major groups of flowering plants. *See* dicot.

exodermis Cylindrical sheet of cells inside the root epidermis of most flowering plants; helps control uptake of water and solutes.

fibrous root system All lateral branchings of adventitious roots arising from a young stem.

flavoprotein Yellow pigment, absorbs blue wavelengths to induce phototropic response.

flower Of angiosperms only, a reproductive structure with nonfertile parts (sepals, petals) and fertile parts (stamens, carpels) attached to a receptacle (modified base of a floral shoot).

fruit [L. after *frui*, to enjoy] Flowering plant's mature ovary, often with accessory structures.

gametophyte (gam-EET-oh-fite) [Gk. *phyton*, plant] Haploid gamete-producing body that forms during plant life cycles.

germination (jur-mih-NAY-shun) Of seeds and spores, resumption of growth after dormancy, dispersal from the parent organism, or both.

gibberellin (JIB-er-ELL-un) Plant hormone; promotes elongation of stems, helps seeds and buds break dormancy, and helps flowering.

gravitropism (GRAV-ih-TROPE-izm) [L. *gravis*, heavy, + Gk. *trepein*, to turn] Directional plant growth in response to gravitational force.

ground tissue system Tissues, parenchyma especially, making up most of the plant body.

growth Of multicelled species, increases in the number, size, and volume of cells. Of bacterial populations, increases in the number of cells.

growth ring One of the alternating bands of early and late wood; a "tree ring."

guard cell One of two cells defining a stoma, an opening across leaf or stem epidermis.

habitat [L. *habitare*, to live in] Place where an organism or species lives; characterized by its physical and chemical features and its species.

hardwood Type of strong, dense wood with many vessels, tracheids, and fibers in xylem.

heartwood A dry tissue at the core of aging stems and roots that no longer transports water and solutes. It helps the tree defy gravity and is a dumping ground for some metabolic wastes.

herbicide Natural or synthetic toxin that can kill plants or inhibit their growth.

homeostasis (HOE-me-oh-STAY-sis) [Gk. *homo*, same, + *stasis*, standing] State in which physical and chemical aspects of internal environment (blood, interstitial fluid) are being maintained within ranges suitable for cell activities.

hormone [Gk. *hormon*, stir up, set in motion] Signaling molecule secreted by one cell that stimulates or inhibits activities of any cell with receptors for it. Animal hormones are picked up and transported by the bloodstream.

humus Decomposing organic matter in soil.

imbibition Water molecules move into a seed, attracted by its hydrophilic groups of proteins.

integrator A control center (e.g., brain) that receives, processes, and stores sensory input, issues commands for coordinated responses.

integument Of animals, protective body cover (e.g., skin). Of seed-bearing plants, one or more layers around an ovule; becomes a seed coat.

internal environment Blood + interstitial fluid.

interstitial fluid (IN-ter-STISH-ul) [L. *interstitus*, to stand in the middle of something] The portion of extracellular fluid that occupies the spaces between animal cells and tissues.

lateral root Outward branching from the first (primary) root of a taproot system.

leaching Removal of some nutrients from soil as water percolates through it.

leaf Chlorophyll-rich plant part adapted for sunlight interception and photosynthesis.

loam Soil with roughly the same proportions of sand, silt, and clay; best for plant growth.

long-day plant Plant that flowers in spring when daylength exceeds a critical value.

magnoliid One of three major flowering plant groups; includes magnolias, avocados, nutmeg.

megaspore Haploid spore; forms by meiosis in the ovary of seed-bearing plants; one of its cellular descendants develops into an egg.

meristem (MARE-ih-stem) [Gk. *meristos*, divisible] One of the localized zones where dividing cells give rise to differentiated cell lineages that form all mature plant tissues.

mesophyll (MEH-zoe-fill) A photosynthetic parenchyma with an abundance of air spaces.

microspore Walled haploid spore; becomes a pollen grain in gymnosperms or angiosperms.

monocot (MON-oh-kot) Monocotyledon; a flowering plant with one cotyledon in seeds, floral parts usually in threes (or multiples of three), and often parallel-veined leaves.

mutualism [L. *mutuus*, reciprocal] Symbiotic interaction that benefits both participants.

mycorrhiza (MY-coe-RIZE-uh) "Fungus-root." A form of mutualism between fungal hyphae and young plant roots. The plant gives up carbohydrates to the fungus, which gives up some of its absorbed mineral ions to the plant.

myelin sheath Lipid-rich membrane formed by oligodendrocytes that wrap around axons of many sensory and motor neurons; enhances long-distance propagation of action potentials.

negative feedback mechanism A homeostatic mechanism by which a condition that changed as a result of some activity triggers a response that reverses the change.

nitrogen fixation Major nitrogen cycle process. Certain bacteria convert gaseous nitrogen to ammonia, which dissolves in their cytoplasm to form ammonium (used in biosynthesis).

nutrient Element with a direct or indirect role in metabolism that no other element fulfills.

organ Body structure with definite form and function that consists of more than one tissue.

organ system Organs interacting chemically, physically, or both in a common task.

ovary (OH-vuh-ree) In most animals, a female gonad. In flowering plants, the enlarged base of one or more carpels.

ovule (OHV-youl) [L. *ovum*, egg] Tissue mass in a plant ovary that becomes a seed; a female gametophyte with egg cell, nutritious tissue, and a jacket that will become a seed coat.

parenchyma (par-ENG-kih-mah) Simple tissue that makes up the bulk of a plant; has roles in photosynthesis, storage, secretion, other tasks.

parthenogenesis (par-THEN-oh-GEN-uh-sis) An unfertilized egg gives rise to an embryo.

perennial [L. *per-*, throughout, + *annus*, year] Plant that lives three or more growing seasons.

pericarp Three tissue divisions of a fleshy fruit; endocarp, mesocarp, and exocarp.

periderm Protective cover that replaces plant epidermis during extensive secondary growth.

phloem (FLOW-um) Plant vascular tissue. Live cells (sieve tubes) interconnect as conducting tubes for sugars and other solutes; companion cells help load solutes into the tubes.

photoperiodism Biological response to change in relative lengths of daylight and darkness.

phototropism Change in the direction of cell movement or growth in response to light (e.g., as in a stem bending toward light).

physiology Study of patterns and processes by which organisms survive and reproduce.

phytochrome A light-sensitive pigment. Its controlled activation and inactivation affect plant hormone activities that govern leaf

expansion, stem branching, stem lengthening, and often seed germination and flowering.

pith Of most dicot stems, ground tissue inside the ring of vascular bundles.

plant physiology Study of adaptations by which plants function in their environment.

plasma (PLAZ-muh) Liquid portion of blood; mainly water in which ions, proteins, sugars, gases, and other substances are dissolved.

pollen grain [L. *pollen*, fine dust] Immature or mature, sperm-bearing male gametophyte of gymnosperms and angiosperms.

pollination Arrival of a pollen grain on the landing platform (stigma) of a flower's carpel.

pollinator Agent (wind, water, an animal) that puts pollen from male on female reproductive parts of flowers of the same plant species.

positive feedback mechanism Homeostatic control; it initiates a chain of events that intensify change from an original condition, then intensification reverses the change.

pressure flow theory Organic compounds flow through phloem in response to pressure and concentration gradients between sources (e.g., leaves) and sinks (e.g., growing parts where they are being used or stored).

primary root First root of a young seedling.

receptor, sensory Sensory cell or specialized structure that can detect a stimulus.

root Plant part, typically belowground, that absorbs water and dissolved minerals, anchors aboveground parts, and often stores food.

root hair Thin extension of a young, specialized root epidermal cell. Increases root surface area for absorbing water and minerals.

root nodule Localized swelling on a root of certain legumes and other plants. Develops when nitrogen-fixing bacteria infect the plant, multiply, and become mutualists with it.

sapwood Of older stems and roots, secondary growth in between the vascular cambium and heartwood; wet, usually pale, not as strong.

sclerenchyma (skler-ENG-kih-mah) Simple plant tissue that supports mature plant parts and commonly protects seeds. Most of its cells have thick, lignin-impregnated walls.

seed Mature ovule with an embryo sporophyte inside and integuments that form a seed coat.

senescence (sen-ESS-cents) [L. *senescere*, to grow old] Processes leading to the natural death of an organism or to parts of it (e.g., leaves).

short-day plant Plant that flowers in late summer or early fall, when daylength is shorter than a critical value.

sieve-tube member One of the cells that join together as phloem's sugar-conducting tubes.

softwood Wood with tracheids but no vessels or fibers. Weaker, less dense than hardwood.

soil Mixture of mineral particles of variable sizes and decomposing organic material; air and water occupy spaces between particles.

sporophyte [Gk. *phyton*, plant] Vegetative body that produces spore-bearing structures; grows by mitotic cell divisions from a plant zygote.

stamen (STAY-mun) A male reproductive part of a flower (e.g., anther with stalked filament).

statolith A gravity-sensing mechanism based on clusters of particles.

stimulus [L. *stimulus*, goad] A specific form of energy (e.g., pressure, light, and heat) that activates a sensory receptor able to detect it.

stoma (STOW-muh), plural **stomata** [Gk. *stoma*, mouth] A gap between two guard cells in leaf or stem epidermis. Opens or closes to control CO_2 movement into a plant and H_2O and O_2 out of it. Stomata help plants conserve water.

surface-to-volume ratio Mathematical relation in which the volume of an object expands in three dimensions (e.g., length, width, depth) but its surface area expands in only two dimensions; a constraint on cell size and shape.

taproot system A primary root together with all of its lateral branchings.

thigmotropism (thig-MOE-truh-pizm) [Gk. *thigm*, touch] Orientation of the direction of growth in response to physical contact with a solid object (e.g., a vine curls around a post).

tissue Of multicelled organisms, a group of cells and intercellular substances that function together in one or more specialized tasks.

tissue culture propagation Inducing a tissue or organism to grow from an isolated cell of a parent tissue growing in a culture medium.

topsoil Uppermost soil layer, the one most essential for plant growth.

toxin Normal metabolic product of a species that can hurt or kill a different species.

tracheid (TRAY-kid) One of two types of cells in xylem that conduct water and mineral ions.

translocation Of cells, movement of a stretch of DNA to a new chromosomal location with no molecular loss. Of vascular plants, distribution of organic compounds by way of phloem.

transpiration Evaporative water loss from a plant's aboveground parts, leaves especially.

turgor pressure (TUR-gore) Type of internal fluid pressure; it acts against a cell wall when water moves into the cell by osmosis.

vascular bundle Array of primary xylem and phloem in multistranded, sheathed cords that thread lengthwise in the ground tissue system.

vascular cambium A lateral meristem that increases stem or root diameter.

vascular cylinder Arrangement of vascular tissues as a central cylinder in roots.

vascular tissue system Xylem and phloem, the conducting tissues that distribute water and solutes through a vascular plant.

vegetative growth A new plant grows from an extension or fragment of its parent.

vein Of a cardiovascular system, any of the large-diameter vessels that lead back to the heart. Of leaves, one of the vascular bundles that thread through photosynthetic tissues.

vernalization Low-temperature stimulation of flowering.

vessel member Type of cell in xylem; dead at maturity, but its wall becomes part of a water-conducting pipeline (a vessel).

xylem (ZYE-lum) [Gk. *xylon*, wood] Of vascular plants, a complex tissue that conducts water and solutes through pipelines of interconnected walls of cells, which are dead at maturity.

ART CREDITS AND ACKNOWLEDGMENTS

This page is an extension of the copyright page. We made every effort to trace ownership of all copyrighted material and to secure permission from copyright holders. Should any question arise concerning the use of any material, we will make necessary corrections in future printings. Many thanks to the illustrators who rendered our art and to authors, publishers, and agents for granting permission to use their material.

Page 487 UNIT INTRODUCTION © 2002 Stuart Westmorland/Stone/Getty Images

CHAPTER 28
28.1 Galen Rowell/Peter Arnold, Inc. **28.2** (a) © K. G. Vock/Okapia/Photo Researchers, Inc.; (b) © Patrick Johns/CORBIS. **28.3** (left above, courtesy Charles Lewallen; left center, © Bruce Iverson; left below, © Bruce Iverson; right, Raychel Ciemma. **28.4** Left above, © CNRI/ SPL/Photo Researchers, Inc.; left below, Dr. Roger Wagner/University of Delaware, www.udel.edu/Biology/Wags; right, art, Lisa Starr with © 2002 PhotoDisc, Inc. **28.5** (a) Gary Head; (b) John W. Merck, Jr., University of Maryland. **28.6** Left © Thomas Mangelsen; right, © Anthony Bannister/Photo Researchers, Inc. **28.7** Left, Giorgio Gwalco/Bruce Coleman; right, © David Parker/SPL/Photo Researchers, Inc. **28.8** © K & K Ammann/Taxi/Getty Images; art, Precision Graphics. **28.9** Left, art, Preface, Inc.; right, photograph, Fred Bruemmer. **28.10** (a) Gary Head with permission from Alex Shigo; (b) From *Tree Anatomy* by Dr. Alex L. Shigo, Shigo Trees and Associates; Durham, New Hampshire. **28.11** (a) www.plants.montara.com; (b) G. J. McKenzie (MGS); (c) Frank B. Salisbury. **28.12** (a) Courtesy of Hall and Bleecker; (b) Juergen Berger, Max Planck Institute for Developmental Biology, Tuebingen, Germany. **28.13** Kevin Somerville and Gary Head. **28.14** Left, © Darrell Gulin/The Image Bank/Getty Images; right, © Pat Johnson Studios Photography. **Page 500** Above, ©PhotoDisc/ Getty Images; below, © Cory Gray. **28.15** (a) Heather Angel; (b) © Biophoto Associates/Photo Researchers, Inc. **28.16** (a) © Geoff Tompkinson/ SPL/Photo Researchers, Inc.; (b) © John Beatty /SPL/Photo Researchers, Inc.

Page 503 UNIT V © Jim Christensen, Fine Art Digital Photographic Images

CHAPTER 29
29.1 (a) R. Barrick/USGS; (b, c) © 1980 Gary Braasch; (d) Don Johnson/Photo Nats, Inc. **29.2** Art, Raychel Ciemma. **29.3** Art, Precision Graphics. **29.4** Art, Raychel Ciemma. **29.5** micrograph, James D. Mauseth, *Plant Anatomy*, Benjamin-Cummings, 1988. **29.6** All, Biophoto Associates. **29.7** (a) D. E. Akin and I. L. Risgby, Richard B. Russel Agricultural Research Center, Agricultural Research Service, U.S. Department of Agriculture, Athens, Georgia; (b) Kingsley R. Stern. **29.8** Art, Precision Graphics. **29.9** George S. Ellmore. **29.10** Art, D. & V. Hennings. **29.11** Art, Raychel Ciemma. **29.12** (a) Robert and Linda Mitchell Photography; (b) Roland R. Dute; (c) Gary Head. **29.13** Art, D. & V. Hennings; (a, left) Ray F. Evert; (a, right) James W. Perry; (b, left) Carolina Biological Supply Company; (b, right) James W. Perry. **29.14** Art, D. & V. Hennings. **29.15** (a) Heather Angel; (b) Gary Head; (c) © 2001 PhotoDisc, Inc. **29.16** (a) Art, Raychel Ciemma; (b) C. E. Jeffree, et al., *Planta*, 172(1):20–37, 1987. Reprinted by permission of C. E. Jeffree and Springer–Verlag; (c) Dr. Jeremy Burgess/SPL/ Photo Researchers, Inc. **29.17** (a) Art, after Salisbury and Ross, *Plant Physiology*, fourth edition, Wadsworth; (b) micrograph, John Limbaugh/Ripon Microslides, Inc.; (c, d) Art, Raychel Ciemma. **29.18** Micrographs, (a) Chuck Brown; (b) Carolina Biological Supply Company. **29.19** Photographs, John Limbaugh/Ripon Microslides; art after T. Rost et al., *Botany: A Brief Introduction to Plant Biology*, second edition, © 1984, John Wiley & Sons, Inc. **29.20** (a) Raychel Ciemma; (b) Alison W. Roberts, University of Rhode Island. **29.21** Art, Raychel Ciemma. **29.22** Art, Precision Graphics. **29.23** John Lotter Gurling/Tom Stack & Associates. **29.24** (a,c) Lisa Starr; (b) H. A. Core, W. A. Cote, and A. C. Day, *Wood Structure and Identification*, second edition, Syracuse University Press, 1979. **29.25** (a) © Jon Pilcher; (b) © George Bernard/SPL/Photo Researchers, Inc.; (c) © Peter Ryan/SPL/Photo Researchers, Inc. **29.26** (a,b) Edward S. Ross. **29.27** Left, NASA/USGS; right, courtesy Professor David W. Stahle, University of Arkansas

CHAPTER 30
30.1 (a,b) Robert and Linda Mitchell Photography; (c) © John N. A. Lott, *Scanning Electron Microscope Study of Green Plants*, St. Louis: C. V. Mosby Company, 1976; (d) Robert C. Simpson/Nature Stock. **30.2** David Cavagnaro/Peter Arnold, Inc. **30.3** (a) William Furgeson; (b) USDA NRCS; (c) U.S. Department of Agriculture. **30.4** (a,c,d) Art, Leonard Morgan; (b) micrograph, Chuck Brown. **30.5** Courtesy Mark Holland, Salisbury University. **30.6** Photographs, (a) Adrian P. Davies/Bruce Coleman; (c) Mark E. Dudley and Sharon R. Long; (a,b) art, Jennifer Wardrip. **30.7** NifTAL Project, University of Hawaii, Maui. **30.8** Micrographs (a) Alison W. Roberts, University of Rhode Island; (b, c) H. A. Cote, W. A. Cote and A. C. Day, *Wood Structure and Identification*, second edition, Syracuse University Press, 1979. **30.9** Left, Natural History Collections, The Ohio Historical Society; Raychel Ciemma. **30.10** Left, M. Ricketts, School of Biological Sciences, The University of Sydney, Australia; inset, Aukland Regional Council. **30.13** © Left, Don Hopey, Pittsburgh Post–Gazette, 2002, all rights reserved. Reprinted with permission; (a,b) Dr. Jeremy Burgess/SPL/Photo Researchers, Inc. **30.14** (a) Courtesy of Professor John Main, PLU; (b) © James D. Mauseth, University of Texas. **30.15** Martin Zimmerman, *Science*, 1961, 133:73–79, © AAAS. **30.16** Left, Palay/Beaubois; right, Precision Graphics. **Page 535** Palay/ Beaubois. **30.18** James T. Brock

CHAPTER 31
31.1 Photographs left, courtesy Merlin D. Tuttle/ Bat Conservation International; right, John Alcock, Arizona State University. **31.2** (a) Robert A. Tyrrell; (b,c) Thomas Eisner, Cornell University. **31.3** Art, Raychel Ciemma and Precision Graphics; photos, Gary Head. **31.4** John Shaw/Bruce Coleman. **31.5** (a) David M. Phillips/Visuals Unlimited; (b) © Dr. Jeremy Burgess/SPL/Photo Researchers, Inc.; (c) David Scharf/Peter Arnold, Inc. **31.6** Raychel Ciemma. **31.7** (a,c,e,f) Michael Clayton, University of Wisconsin; (b) Raychel Ciemma; (d) Dr. Charles Good, Ohio State University, Lima. **31.8** (a) Dr. Dan Legard, University of Florida GCREC, 2000; (b) Richard H. Gross; (c) © Andrew Syred/SPL/ Photo Researchers, Inc.; (e) Mark Rieger; (f–h) Janet Jones. **31.9** (a) R. Carr; (b) Gary Head; (c) Rein/Zefa. **31.10** John Alcock, Arizona State University. **31.11** Russell Kaye, © 1993 The Walt Disney Co. Reprinted with permission of *Discover* Magazine. **31.12** (a) Runk & Schoenberger /Grant Heilman Photography, Inc.; (b) Kingsley R. Stern. **Page 548** Raychel Ciemma. **31.13** Gary Head

CHAPTER 32
32.1 (a) Michael A. Keller/FPG; b,c) R. Lyons/ Visuals Unlimited. **32.2** Sylvan H. Wittwer/ Visuals Unlimited. **32.3** © Dr. John D. Cunningham /Visuals Unlimited. **32.4** (a,b) Raychel Ciemma; (c) Herve Chaumeton/Agence Nature. **32.5** (a,b) Raychel Ciemma; (c) Barry L. Runk/Grant Heilman Photography, Inc.; (d) from Mauseth. **32.6** (a,b) Raychel Ciemma; (c) Biophot. **32.7** (a) Kingsley R. Stern; (b) Precision Graphics. **32.8** (a) Michael Clayton, University of Wisconsin; (b) John Digby and Richard Firn. **32.9** (a,b) Micrographs courtesy of Randy Moore, from "How Roots Respond to Gravity," M. L. Evans, R. Moore, and K. Hasenstein, *Scientific American*, December 1986. **32.10** (a) © Adam Hart-Davis/SPL/Photo Researchers, Inc.; (b,c) Lisa Starr. **32.11** Gary Head. **32.12** Cary Mitchell. **32.14** Precision Graphics. **32.15** Long-day plant, © Clay Perry/ CORBIS; short-day plant, © Eric Chrichton/ CORBIS; art, Gary Head. **32.16** (a) Jan Zeevart; (b) Ray Evert, University of Wisconsin. **32.17** Above, N. R. Lersten; below, © Peter Smithers/CORBIS. **32.18** Larry D. Nooden. **32.19** R. J. Down. **32.20** Art, Lisa Starr; photograph, Eric Welzel/Fox Hill Nursery, Freeport, Maine. **32.21** Gary Head and Preface Graphics Inc. **32.22** Inga Spence/ Tom Stack & Associates. **32.23** Grant Heilman Photography, Inc.

INDEX   *The letter* i *designates illustration;* t *designates table.*

A

ABA, 530, 551, 554t, 555
ABC model, flowering, 498–499
Abscisic acid, 530, 551, 554t, 555
Abscission, 560, 560i
Absorption, heat
  by yellow bush lupine, 496
Absorption, light
  in photosynthesis. *See* Photosynthesis
  by phytochrome, 558, 558i
Absorption, nutrients
  by carnivorous plants, 522, 522i
  by roots, 524, 526–527, 526i
Absorption, water
  plants, 523, 524, 526–527, 527i
  seed germination and, 552
Accessory fruits, 542t, 543, 543i
Acer (maple), 512i, 544, 544i
Acorn, 542t, 543
Active transport
  defined, 501
  key role in metabolism, 501
  neural function and, 501
  in solute–water balance, 501
  stomatal function and, 531
Adaptation
  defined, 489, 492
  difficulty identifying, 493, 493i
  in gas exchange, 491, 493, 500
  internal transport, 500
  llama–camel question, 493, 493i
  long–term, 489, 492–493
  natural selection and, 492
  nature of, 489, 490–493, 493i
  plants, 501
  salt–tolerant tomatoes, 492, 492i
  short–term, 492
Adventitious root, 514
Aesculus, 512i
Aflatoxin, 562
African violet, 546, 546t
Agapanthus, 512i
Agar, 555i
Agent Orange, 555
Aggregate fruit, 542t, 543
Aging
  of trees, 496
Agriculture
  erosion and, 525
  pesticides, 562
  plant hormones in, 550–551, 555, 562
  salinization with, 492, 492i
  soils and, 524–525, 525i
Air pollution
  stomata and, 531, 531i
Alachlor, 562i
Alar, 562i
Alder, 504
Alfalfa, 510, 511i, 516, 527
Allergic rhinitis, 539
Allergy
  pesticides and, 562
  pollen and, 539
Almond, 542t
Aluminosilicate, 524
Ananas, 543, 543i
Anatomy, defined, 489
Andean goose, 488
Andes, 488, 492
Angiosperm. *See also* Flowering plant
  characteristics, 505
  life cycles, 539, 540, 540i
Angraecum sesquipedale, 537
Animal tissue
  defined, 490
Annual, 516
Anser indicus, 488, 488i, 489i
Anther, 538i, 539, 540, 540i
Anthocyanin, 538
Aphid, 532, 532i

Apical dominance, 555
Apical meristem, 505, 554t, 555
  primary root, 514, 514i, 552i
  primary shoot, 507i, 510, 510i, 552i
Apple, 542t, 543, 543i
Apricot, 542t, 561
Arabidopsis thaliana, 498, 498i,
  558, 564
Armillaria, 496
Arsenic, 562
Asthma, 562
Atrazine, 562i
Attalea, 512
Autoimmune response
  examples, 499
Auxin, 551, 554t, 555–557, 556i,
  560, 562
Axon
  in multiple sclerosis, 499, 499i
  structure, 499i

B

Bahama woodstar, 537i
Bald cypress, 521i
Banana, 542t
Barheaded goose, 488, 489i
Bark, 518, 518i, 519, 519i
Bat
  pollinators, 536i, 537, 545
Bean plant, 497i, 509, 542t, 552,
  552i–553i
Bedstraw, 544
Bee
  pollinators, 537, 537i
Bee, D., 549
Beetle
  pollinators, 536–537
Bermuda grass, 546t
Berry, 542t, 543, 543i
Biennial, 506, 516, 561
Bill
  bird, 537i
Biological clock
  defined, 558, 563
  in plants, 558–559, 558i, 559i
Bird
  pollinators, 536, 536i, 545
Black locust, 512i
Blackberry, 504, 542t
Blade
  leaf, 512, 512i, 564
Body temperature, 494–495
Bog, 522, 525
Boron plant nutrient, 524t
Bottlenose dolphin, 487i
Bradyrhizobium, 527i
Brain
  homeostasis and, 494, 494i
Brassinolides, 554t
Bristlecone pine, 519
Bud
  defined, 510
  dormancy, 554t, 555
  lateral (axillary), 506i, 510i, 512i, 555
  terminal, 506i, 516i
Bud scale, 510
Bulk flow
  in phloem, 533i
Buttercup, 515i
Butterfly
  pollinators, 537

C

C4 plant, 531
Cabbage, 550, 551, 551i
Cacao, 544, 544i
Cactus, 504, 536i
  C4 pathway in, 531
  camouflaged species, 501i

Calcium
  dietary, 524t
  in plant function, 524, 524t, 530, 558
California poppy, 514i, 550i
Callus, 546
Calyx, 538, 538i, 539i, 543
CAM plant, 531
Camel, 493, 493i
Camellia, 512
Capillary
  pulmonary (lung), 491, 491i
  structure, 491i, 501i
Capillary bed
  exchanges, 500
  function, 500
Capsella, 542, 542i
Carbohydrate
  storage in plants, 532
Carbon
  plant growth and, 524, 524t
Carbon dioxide
  atmosphere, 523
  in photosynthesis, 518
  uptake by plants, 513, 513i, 518, 522,
    530–531, 530i, 531i
Carcinogen, 555
Carica papaya, 549, 549i
Carnivorous plant, 522, 522i, 523i
Carotenoid, 538
Carpel, 498i, 499, 538i, 539, 540i
Carrot, 514, 516, 542t
  tissue culture propagation, 546, 547i
Casparian strip, 526, 526i
Castanea, 516
Cattle
  grazing, 535
  hormone–induced growth of, 564
Celery, 550, 551, 562
Cell communication
  in *Arabidopsis*, 498–499, 498i, 558
  flowering and, 498–499, 558–559
  gene activity and, 550
  hormones in, 550, 551
  nature of, 489, 498–499, 552
  in nervous system, 499
  in plant development, 498, 498i,
    550–559
  receptors, 498, 552
  signal reception, transduction, and
    responses, 498–499, 554
Cell differentiation
  plants, 505, 507i, 510, 514i, 538, 548,
    551, 553
Cell division
  meristematic, 505, 507, 507i, 510,
    512i, 514–515, 542, 548, 552, 554
Cell wall
  plant, 508, 508i, 515, 526, 528, 554
Cellulose
  digestion, 560
  plant cell, 508, 527i, 530, 554i
  in root nodule formation, 527i
Cherry (*Prunus*), 538, 542t
  life cycle, 538, 538i, 540i
Chestnut, 516
Chickweed pollen, 539i
Chimpanzee
  sensory receptors, 494, 494i
Chlordane, 562i
Chlorine
  plant growth and, 524t
Chlorophyll
  synthesis, 524t
Chloroplast
  guard cell, 531i
  plant, 513
Chlorosis, 524t
Chocolate, 544i
Chrysanthemum, 558, 559, 559i
Cilium (cilia)
  in airways, 491i

Circadian rhythm, 497, 497i, 558
Clay, 524, 525
Cloephaga melanoptera, 488
Clone
  aspen, 546, 547i
Cloning
  plants, 546–547, 547i
Coast redwood, 516, 518i
Cobra lily, 523i
Cocklebur, 544, 558, 559
Coconut, 544, 546
Coconut palm, 544
Coevolution
  defined, 536
  plant/pollinator, 536, 536i
Cohesion–tension theory, 528–529,
  529i, 535
Coleoptile, 552i, 553i, 555, 555i,
  556, 557i
Coleus, 510i
Collenchyma, 506, 508, 508i, 509, 520t
Columbus, C., 543i
Community
  cellular, 490, 498
Companion cell, 508, 532, 532i
Compartmentalization
  woody plants, 496, 496i
Competition, intraspecific
  parent plants/seedlings, 544
Complex plant tissue, 508–509, 508i,
  509i, 520t
Compound leaf, 512, 512i
Concentration gradient
  in animal function, 500, 501
  diffusion and, 500
  factors affecting, 500
  gas exchange and, 500, 508
  neural membrane, 499i
  in plant function, 500, 501, 531, 532
  in solute–water balance, 501
Conifer
  vascular tissues, 509, 511i
Contraction
  cardiac. *See* Cardiac muscle
  labor (birth process), 495
Copper, 524t
Cork, 509, 518, 519i, 520t
Cork cambium, 520t
  activity at, 518, 519i
  meristematic source, 507, 507i
Corm, 546t
Corn (*Zea mays*)
  body plan, 506, 509
  crops, 544, 547, 562
  development, 542, 552, 552i
  kernel (grain), 542, 542t, 543, 552,
    552i, 556i
  leaf, 512, 512i
  root, 514i, 515i, 526, 556i
  root cap, 514i, 556i
  seed, 542, 544
  stem, 511i
  tissue culture propagation, 546t, 547
  water demands by, 526
Corolla, 538, 539i
Cortex
  root, 515, 515i, 517i
  stem, 510, 511i
  wood formation and, 518, 519i
Cotyledon, 509, 542–543, 542i, 543i,
  552, 552i
  dicot vs monocot, 509, 509i, 542
Crain, Richard, 558
Creosote, 546
Crop, plant
  dispersals, 544
  dormancy requirements, 561
  propagation, 546–547
Crop dusting, 562i
Crop production
  commercial tomatoes, 492, 492i

fertilizers and. *See* Fertilizer
pesticides and, 562, 562i
plant hormones and, 550–551
salt tolerance and, 492, 562
toxic metals and, 562
Cucumber, 542t
Cuticle, plant, 513, 530–531
function, 509, 523
micrographs, 509i, 530i, 531i
Cutin, 530
Cytokinin, 551, 554t, 555, 558

D

2,4–D, 555, 562i
Daisy, 537
Daminozide, 562i
Dandelion, 514, 544
Darlingtonia californica, 523i
Darwin, C.
on phototropism, 556
Daucus carota, 546, 547i
Day–neutral plant, 558
Daylength
plants and, 558–561, 558i, 559i, 561i
DDT, 562, 562i
Death
plant, 560–561
Deciduous plant, 512
Decomposer
fungal, 496
soil properties and, 524
Dehiscent fruit, 542t
Dendroclimatology, 521
Dermal tissue system, plant, 506, 506i, 509, 510, 510i, 514, 514i, 517i
Development
controls, 552
defined, 490, 554
growth vs, 554
Development, plant
dicot vs monocot, 552, 552i
early patterns, 552–553, 552i
environment and, 552–553, 556–561
fruit formation in, 542, 543, 543i, 545
hormones in, 550–555
primary growth, 505, 507, 510i, 514, 514i, 515i, 516i
secondary growth, 505, 506, 507, 516–517, 516i, 517i, 518, 519, 519i
seed formation in, 542–543, 542i
Dicot (Eudicot), 506–507, 506i, 507i, 509i, 514, 520i
development, 552–553, 552i
growth, 552–553, 552i
monocot vs, 509–511, 509i, 511i, 512, 512i
Dictyostelium, 498
Diffusion, 500
capillary bed, 500
factors affecting, 500
in plants, 500, 512, 513, 513i, 530–531
at respiratory surfaces, 500
Digestion, carnivorous plant, 522, 522i
Digestive enzyme
seeds and, 542, 544, 545
Dionaea muscipula, 522–523, 522i
Dioxin, 555
Division of labor, 490, 494
Dixon, H., 528
Dormancy, 554, 555, 560–561, 561i
Double fertilization, 540i, 541, 541i
Douglas fir, 504, 504i, 505i, 560–561, 561i
Dove, 545
Dromedary, 493, 493i
Drought
Virginia colonists and, 521, 521i
Drupe, 542t
Dry fruit, 542t, 543
Duckweed, 512

E

Early wood, 519
Ecological succession
after volcanic eruption, 504–505
Edelweiss, 488, 497
Effector, for nervous systems, 494, 494i, 495i
Elm tree, 519i
Embryo sac, dicot, 540, 540i
Embryo sporophyte, 540i, 542, 542i, 544, 552i
development, 552, 552i, 554t
Endocarp, 543
Endocrine system
links with nervous system, 495
Endodermis, 526, 526i
Endosperm, 540, 540i, 541, 542i, 552, 552i
monocot vs dicot, 542
Energy
sunlight, 512
Environment
effects on wood formation, 518–519
in natural selection. *See* Natural selection
plant responses to, 552–555, 558–559
pollution of. *See* Pollution
Enzyme
plant, 553, 554
Eocene, 493
Epidermis, plant
defined, 506, 506i
function, 506, 508
leaf, 508, 508i, 513, 513i
meristematic source, 507, 507i
root, 514, 514i, 515i, 517i, 526, 526i, 527i
shoot, 510, 510i
specializations, 508, 513, 522, 530–531, 531i
Erethizon dorsatum, 501i
Erosion, 525, 525i
Eschscholzia californica, 514i, 550i
Ethylene, 551, 554i, 554t, 555
Eudicot, 506, 509, 509i
Evaporative water loss, 528, 529i
Evergreen, 512
Exocarp, 543
Exodermis, 526, 526i
Experiment, examples. *See also* Test, observational
aphids and phloem, 532
auxin effects, 551i, 555, 555i
circadian rhythm, plants, 558, 559i
dormancy, fir *seedlings*, 560, 560i
dormancy, lilac, 561i
flower/set fruit ratio, 545
flowering response, 558–559, 559i
foolish *seedling* effect, 551, 551i
gibberellin effects, 554–555, 555i
mechanical stress on plant, 557, 557i
nitrogen deficiency, on plant growth, 527i
plant hormones, 550–551, 550i, 551i
pressure flow theory, 533
senescence in plants, 560i
stomatal action, 531i
tissue culture propagation, 546–547
vernalization, 561, 561i
wilting, tomato plants, 492, 492i
Extracellular digestion and absorption by carnivorous plants, 522
Extracellular fluid
defined, 494

F

Farmland erosion, 525
Fat (lipid)
metabolism, 532
storage, 532
Feedback mechanism, 494–495, 495i
homeostasis and, 494–495, 495i

negative, 494
positive, 495
Fertilization
double, 540i, 541
plants, 538–541, 538i, 540i, 541
Fertilizer
in soils, 523, 525
Fiber
plant, 502, 508, 508i, 511i
Fibrous root system, 514
Fig, 542t
Fir, 560, 561i
Fireweed, 504
Flatworm
characteristics, 500, 500i
Flavoprotein, 557
Flax, 508i
Fleshy fruit, 542t, 543
Floral spur, 537
Floral tube, 536, 537, 537i
Flower
characteristics, 506, 506i, 524t, 536–539, 538i, 539i
color, 536–537, 537i, 538
components, 538–539, 538i, 539i
defined, 506, 536
development, 498–499
examples, 520i
fruit set vs number of, 545, 545i
function, 536i, 537, 538, 545
in life cycles, 540i
nectar guide, 536–537, 537i
perfect vs imperfect, 539
structure, 536
Flowering plant. *See also* Angiosperm
asexual reproduction, 546–547, 546t, 547i
biodiversity, 505, 506
biological clocks, 558–559, 558i, 559i
carnivorous, 522–523, 522i, 523i
characteristics, 505, 506, 506i, 512, 519, 520t
coevolution, pollinators, 536–537
development, 540–543, 540i–543i
dicot vs monocot, 509, 509i
evolution, 536, 545
hormones. *See* Hormones, plant
nature of growth, 507, 507i, 554–555, 554t
oldest known, 546
primary growth, 506–515, 506i–515i, 538
root, 507i, 514–515, 514i
secondary growth, 506, 506i, 516–517, 516i, 517i
sexual reproduction, 538–541, 538i–541i
shoot, 506–511, 506i–511i
tissues, 506, 508–509, 508i, 509i, 520t
Flowering response, 554t, 555, 558–559, 558i, 559i
Fly
pollinators, 537
Venus flytrap and, 522, 522i
Foolish seedling effect, 551, 551i
Forest
aspen, 546, 547i
Four-leaf clover, 512i
Fragaria, 543, 543i
Freeway landscape plants, bad choices, 559
Fruit
categories, 542t
defined, 542
dispersal, 544
examples, 543i, 544, 546t
formation, 542–543, 543i, 545, 545i, 549
function, 537, 544
hormones and, 554–555, 554t
ripening, 554t, 555
Fungicide, 562
Fungus (Fungi)
crop–destroying, 550

of mycorrhizae. *See also* Mycorrhiza
in orchid germination, 547
pathogenic, 496, 496i, 550
saprobic, 496, 496i, 527
symbionts, 527
toxic, 562
Fusiform initial, 516–517, 516i

G

Galápagos tomato, 492, 492i
Gamete formation
plants, 540, 540i
Gametophyte
defined, 538, 538i, 539, 539i, 548
female, 538
flowering plant, 538–541, 538i, 540i, 541i
male, 538
seed plants, 538–539, 538i
Gardenia, 555i
Gas exchange. *See also* Respiration
leaf adaptations for, 512–513
in mesophyll, 508, 512
Gassner, G., 561
Gene. *See also specific types*
master, 498–499, 558
Gene control, eukaryotic
in plant development, 553–554
Gene expression
environment and, 552
hormones and, 544
in plant development, 552–553
selective. *See* Selective gene expression
Genetic divergence
in adaptive radiations. *See* Adaptive radiation
llamas/camels, 493
Genetic engineering
crop plants, 562
Germination, 524t, 542
defined, 538, 548, 552
orchid, 547, 547i
pollen grain, 539, 540
seed, 500, 505, 509, 514, 516, 541, 552, 552i, 555, 558, 560
Giant saguaro, 536i, 545, 545i
Gibberella fujikuroi, 550
Gibberellin, 550–551, 550i, 551i, 554i, 554t, 555, 560, 561
Girdling, 521
Gladiolus, 546t
Gleditsia, 512
Glucose
how plants use, 532
Glyphosate, 562i
Goose
high–altitude, 488, 488i, 489i
Graft
plant, 546
Grain, 542t
Grant, M., 547i
Grape, 542t, 543, 550, 550i, 551
gibberellin and, 550i, 551
Grapevine, 546
Grass, 539i
pollen, 539i
Gravitropism, 554t, 556, 556i
Ground meristem, 506i, 507, 507i
Ground tissue, plant
characteristics, 506, 506i, 509, 509i, 510, 511i, 513, 514–515
defined, 506
embryonic source, 507, 507i
of flowers, 538
functions, 506, 507
of stems, 510–511, 511i
Growth
cell divisions in. *See* Cell division
defined, 490
development vs, 554
Growth, plant
defined, 554

environment and, 552–553, 556–562
hormones in. *See* Hormones, plant
inhibitors, 555, 556, 557, 557i, 558i
nature of, 507, 507i
nutrients for, 524–525, 524t
patterns, 552–553, 557
primary, 507–515, 507i–515i, 516, 516i
processes, summary, 534
secondary, 507, 507i, 516–519,
  516i–518i
soils and, 525
tropisms and, 556–557, 556i, 557i
vegetative, 546, 546t
Growth hormone, 564. *See also*
Growth ring, 519, 519i, 521, 532i
Guard cell, 509, 513i, 520t, 527
  530i, 531i
Gymnosperm
  defined, 505
  roots, 527

H

Habitat
  defined, 501
Hakea gibbosa, 530i
Hardwood, 519, 519i
Hawkmoth, 537
Hay fever, 539, 540
Heartwood, 518, 519i
Helianthus, 508i, 542t, 556i, 564i
Hemlock, 504
Hemoglobin
  high altitude and, 488, 493, 493i
Herbaceous plant, 516
Herbicides, 555, 562, 562i
Hesperidium, 542t
Hibiscus, 537i
Hickory, 519
Himalayas, 488, 489i
Hives (allergic reaction), 562
Holly, 531i
Homeostasis
  in animals, 494–495, 494i, 495i
  defined, 489, 491
  maintenance of, 494–495, 494i, 495i
  in plants, 496–497, 496i, 497i
Honey locust, 512i
Honeydew, 532, 532i
Hormones, animal
  signaling mechanisms, 498
Hormones, plant
  abscisic acid, 554t, 555, 561, 563
  auxins, 551, 554t, 555, 555i, 556, 556i,
    557, 560, 562, 562i, 563
  cytokinin, 554t, 555, 560, 563
  defined, 550
  ethylene, 554t, 555, 557, 560, 563
  gibberellins, 550–551, 550i, 551i, 554t,
    555, 560, 561, 563
  main categories, 551, 554t, 555
  in parthenogenesis, 546
  stomatal function and, 531i
Houseplant, propagating, 546, 546t
Human
  plants and, 544, 544i
Human body
  respiratory system, 490i
Human history
  Roanoke Island colonists, 521
Hummingbird, 537i
  pollinator, 537i
Humus, 525
Husky, 494–495, 495i
Hybridization
  orchids, 547i
Hydrogen
  ionized. *See* Hydrogen ion
  in organisms, 524t
Hydrogen bonding
  transpiration and, 528, 529, 529i
Hydrogen ion (H+)
  in soil ion exchanges, 524

Hypericum, 520i
Hypha, 527
Hypocotyl, 552i
Hypotonic solution
  soilwater and plant growth, 524i

I

IAA, 555, 555i, 560
Imbibition, 552
Imperfect flower, 539
Imperial American macaw, 489i
Indehiscent fruit, 542t
Indoleacetic acid, 555, 555i
Induced propagation, 546–547
Industrial smog, 531i
Infections, fungal agents
  Armillaria, 496, 496i
Inheritance
  plants, 552–553
Insect
  as food competitors with
    humans, 562
  as pollinator, 536, 536i, 545
Insecticide, 562, 562i
Integrator, 494, 494i, 495i
Integument, 540
  ovule's, 540, 541i, 542
Internal environment
  defined, 491
  homeostasis and, 494–495, 502
  nature of, 500
Internal transport, 500
Internode, 506i, 510
Interstitial fluid, 494
Ion exchange
  in soil, 524–525
Iris, monocot, 520i
Iron
  plant nutrient, 524t

J

Jade plant, 546t
Jasmonates, 554t
Juglans, 516i

K

Kaffir lily, 509i
Kissing, 494, 494i
Kurosawa, E., 550

L

Labor contraction, 495
Larkspur, 542t
Larrea, 546
Lasso, 562i
Late wood, 519
Lateral bud, 506i, 510, 510i
Lateral meristem, 507, 507i, 516.
  See also Cork cambium;
  Vascular cambium
Lateral root, 500i
  fibrous root system, 514, 514i
  formation, 514, 515, 515i
  taproot system, 514, 514i
Leaching, 525, 525i
Lead
  as insecticide, 562
Leaf
  air pollution and, 531, 531i
  characteristics, 500, 501i, 506, 506i,
    509i, 512–513, 512i
  development, 552i
  diffusion and, 500
  epidermis, 509i, 513, 513i
  fine structure, 512, 513, 513i
  growth, 512, 524t, 554, 558
  mesophyll, 513, 513i
  monocot vs dicot, 509i, 512, 512i
  photosynthesis in. *See* Photosynthesis

rhythmic movements, 558
  specializations, 512, 512i, 522, 522i,
    523i, 530
  stomata, 509, 513, 513i, 550–551,
    550i, 551i
Leafy gene, 499
Legume, 542t
  root nodules, 527, 527i
Lemna, 512
Lemon, 542t
Lenticel, 518
Leontopodium alpinum, 488, 489i
Leukemias, 555
Life cycle, examples
  dicot (*Prunus*), 538–539, 538i, 540i
  flowering plant, 516, 538, 538i,
    540i–541i, 551, 560–561
  plant, generalized, 560–561,
    560i, 561i
  seed–bearing plant, 552
Lignin
  characteristics, 508, 508i, 509
Lilac, 561i
Lilium, 509, 509i, 523i, 546t
Lily pad, 500, 500i
Liver
  disorders, 554
Llama, 493, 493i
Loam, 525
Long–day plant, 558–559, 559i
Lost colony, Roanoke Island, 521, 521i
Luna, old–growth redwood, 521
Lung
  function, 500. *See also* Gas exchange
  human, 491i
Lupinus arboreus, 497, 497i
Lycopersicon, 490i, 492, 492i, 506i,
  542t, 543, 557i

M

Macronutrient, plants, 524, 524t
Madagascar hawkmoth, 537
Magnesium
  plant, 524t
Magnoliid, 506
Malathion, 562i
Malus, 543, 543i
Manganese, 524t
Maple, 509, 512i, 518
  fruit, 542t, 544, 544i
  leaf, 560i
Maple syrup, 518
Marigold, 516, 537i
Marsh marigold, 537i
Master gene
  Arabidopsis thaliana, 498–499, 558
Medicago, 511i, 516
Megaspore, 537
  dicot life cycle, 540, 540i
Mercury, used as insecticide, 562
Meristem
  apical. *See* Apical meristem
  callus, 546
  defined, 498, 507
  lateral. *See* Lateral meristem
  nature of, 498, 499, 506, 507, 507i
  orchid, 547i
Mesocarp, 543
Mesophyll, 508, 532
  palisade, 513, 513i, 530i
  spongy, 513, 513i
  Venus flytrap, 522i
Mesquite, root system, 515
Metabolism, plant
  nutrients required for, 524, 524t
  photosynthesis. *See* Photosynthesis
  translocation and, 532
Microfibril, cellulose, 554i
Micronutrient, plant, 524, 524t
Microspore
  dicot life cycle, 540, 540i
  pine life cycle, 536i, 537

Microtubule
  plant cell, 555i
Mineral
  in plant nutrition, 524, 524t
Mitotic cell division
  plants, 505, 538–540, 540i, 546, 551
  time of, life cycles. *See also* Life
    cycle, examples
Molybdenum, 524t
Monocot
  characteristics, 506, 509, 512i, 513
  dicot vs, 509–511, 509i, 511i,
    512, 512i
Moth, 537
Mount Everest, 488, 488i
Mount Saint Helens, 504–505,
  504i, 505i
Mountain climbing, 488, 488i
Mucigel, 514i
Multiple fruit, 542t, 543
Multiple sclerosis, 499, 499i
Mustard, 542t
Mutation
  Arabidopsis thaliana, 498, 564
  beneficial, 547
  effects on development, 493
  evolution and, 488
  natural selection and. *See* Natural
    selection
  neutral, 493
  in tissue culture propagation, 547
Mutualism. *See also* Symbiosis
  bacterium/root, 527, 527i
  defined, 527
  legumes/nitrogen–fixing bacteria,
    527, 527i
  mycorrhiza, 527
  plant/pollinator, 536–537
Mycorrhiza, 527
Myelin, defined, 499
Myelin sheath
  in multiple sclerosis, 499, 499i
  in signal propagation, 499, 499i

N

Natural selection
  of coevolved species, 536
  of xylem components, 528i
Nectar, 536–537, 536i, 537i, 545
Nectar guide, 537, 538
Negative feedback
  defined, 494
  examples of, 494–495, 494i
Nelumbo nucifera, 503i
Neuron
  pesticide effects on, 562, 562i
  sensory. *See* Sensory neuron
Nitrogen
  deficiency, 522, 527, 527i, 535
  metabolism, 524t
  plant growth and, 524, 524t, 535
Nitrogen fixation, 527
Nitrogen–fixing bacterium,
  527, 527i
Node
  plant stem, 507i, 510, 512i, 552i
Nonwoody plant, 516
Nucleic acid
  plant functioning, 524t
Nursery stock
  dormancy in, 561
  hormone applications to, 555
Nutrient
  plants, 522–525, 524t, 527, 527i, 535
Nutrition
  plant, 522–525, 542

O

Oak
  flowers, 539
  hardwood, 519, 519i

leaf, 512i
root system, 514
Observational test, examples
gravitropism, 556i
phototropism, 557i
rhythmic leaf movements, 497, 497i
root system growth, 514
Oil spills, 518
Olive, 542t
Onion, 546t
Orange, 542t, 544, 546, 546t, 551
Orchard, 555
Orchid, 509, 537, 544
commercial propagation, 546t, 547, 547i
Organ, animal
defined, 490
Organ formation
in plants, 498–499
Organ identity gene, 498
Organ system. *See also specific types*
defined, 490
functions, 491
Organic compound
distribution in plants, 523, 532–533
in topsoil, 525i
toxic, 562
Organophosphate, 562i
Oryx, 492i
Osmosis
guard cell function and, 530
in translocation, 533, 533i
wilting and, 530
Ovary, flowering plant
defined, 539
in fruit formation, 542–543
function, 539–541
Ovule
flowering plant, 536
formation, 537, 540, 540i
seed development from, 536, 536i
Oxygen
aerobic respiration and, 518
at high altitudes, 493, 493i
leaf adaptations to, 512, 513i
plant growth and, 518, 524, 524t
plant nutrient, 524t
Oxytocin, 495

P

Paleo–Indian dispersals, 493
Palisade mesophyll, 513, 513i
Palm, 509, 512
Pando the Trembling Giant, 547i
Papaya, 549, 549i
Parenchyma
characteristics, 506, 508, 509, 520t
functions, 508, 513, 518, 532
micrographs, 508i, 509i, 513i
photosynthetic, 508, 513, 513i
in secondary growth, 518
Parenchyma cell, 508, 509i
Parthenogenesis, 537, 546, 546t
Passiflora (passion flower), 557i
Passive transport
nature of, 501
Pea, 527, 542t, 543
Peach, 542t, 543, 561
Pear, 508i, 542t
Pectin, 508, 530, 560
Pepo, 542t
Perennial
defined, 516
in life cycles, 560, 561, 561i
Perfect flower, 539
Perforation plate, 528i
Pericarp, 543, 544, 552i
Pericycle
in lateral root formation, 515, 515i
location, 514i, 515, 515i, 517i
meristematic source, 514i, 515, 517i
Periderm, 518, 520t

Pesticide
effects on humans, 562
natural, 562
toxic, 562, 562i
Petal, 499, 537, 538, 538i, 539i, 543i
Petiole, 512, 512i, 560, 560i
Petunia, 549
Pfr, 558, 558i, 560, 561i
Phaseolus, 497i, 513i, 552i–553i
Phenol, as plant defense, 496, 554t
Phloem
components, 505, 508, 509, 509i, 520, 520t, 532, 533i
defined, 506, 508
function, 509, 521, 523, 532
primary, 510, 510i, 514i, 515, 515i, 526i
secondary, 507i, 516–518, 516i–518i
in translocation, 532–533, 533i
Phospholipid
characteristics, 524t
Phosphorus
plant nutrient, 524t
Photoautotroph
algal. *See* Alga
plant, 522
Photolysis, 524t
Photoperiodism, 558–559, 559i
Photosynthesis
in cotyledons, 552i
environment variations and, 531
leaf adaptations for, 512, 513, 530–531
nutritional requirements, 524t
plant growth and, 534i
process, 531
products, 532
rates, 530
translocation and, 532, 533i
Photosynthetic cell. *See also* Photoautotroph
in leaf, 513, 513i, 529i, 531i, 533i
parenchymal, 508, 513
Phototropism, 554t, 556–557, 557i
Physiology, 489, 522
Phytochrome
in biological clocks, 558, 558i, 559, 563
in leaf folding, 497
Pigment
floral, 536, 537, 538
in phototropism, 556–557, 557i
phytochrome, 497, 558, 558i
plant, 558, 558i
Pine (*Pinus*)
farming, 519i
wood, uses of, 519i
Pineapple, 542t, 543, 543i
Pinus, 519i
Pistil, 539. *See also* Carpel
Pith, plant, 510, 510i, 511i, 515, 515i
Plant. *See also specific types*
annual vs biennial, 516
carnivorous, 522, 522i, 523i
classification. *See also* Appendix I
clones, 546
defenses, 508, 518, 519
development. *See* Development, plant
evolution, 536, 545
growth. *See* Growth, plant
homeostasis in, 496–497
hormones. *See* Hormones, plant
life cycles. *See* Life cycle, examples
master genes, 498–499, 498i
number of species, 506
nutrient uptake by, 523–524, 526–527
nutrition, 522–525, 524t, 527,527i, 535
propagation, 546–547
reproduction. *See* Plant reproduction
specimen cuts from, 507i
tissue. *See* Plant tissue
translocation, 532–533, 533i
"typical," 506, 506i

vascular, 500, 505, 506ff, 523
water conservation, 530–531
water transport, 500, 501i, 526
water uptake by, 523, 525–527, 528i, 529i
woody vs nonwoody, 516
Plant cell
categories, 506, 508–509
growth, 554, 554i
meristematic, 507, 507i, 511i, 552
metabolism. *See* Photosynthesis
photosynthetic, 508, 509i, 512–513
primary wall, 554, 554i
secondary wall (lignified), 508, 508i, 509, 519
shape, 554, 554i
Plant growth
during early development, 552–553, 552i–553i
flowering, 558–559, 558i, 559i
growth rates, 556–557, 556i, 557i
hormones and. *See* Hormones, plant
in life cycles, 560–561, 560i, 561i
mechanical stress and, 557, 557i
primary, 507, 507i, 510–511, 510i, 511i, 514–515, 514i, 515i, 552–554, 557, 560
secondary, 507, 507i, 516–519, 516i–517i
Plant physiology, defined, 522
Plant reproduction
asexual, 537, 546–547, 546t, 547i
during life cycle. *See* Life cycle, examples
flowering plants, 536–541, 536i
representative dicot, 540, 540i
sexual, 536–541
Plant tissue
complex, 505, 506, 508–509, 509i
leaves, 512–513, 513i
major systems, 505, 506
plant body, 506–507, 506i, 507i
primary, 505–515, 552, 557, 560
roots, 514–515, 514i, 515i
secondary, 507, 507i, 516–519
shoots, 510–511, 510i, 511i
simple, 505, 506, 508, 508i
summary of, 520t
types, 490, 490i, 508–509, 508i, 509i
wood and bark, 518–519, 518i, 519i
Plasma, 494
Plastid, 532, 556
Poinsettia, 558, 559
Polar bear, 492, 492i
Pollen grain
allergies and, 539, 539i
components, 539i, 540
dicot vs monocot, 509i
dispersal, 536, 536i
flowering plant, 540, 540i, 545
formation, 540–541, 540i–541i
structure, 539i
Pollen sac, 539, 540i
Pollen tube, 540, 540i–541i
Pollination
defined, 540
flowering plant, 536–537, 536i, 537i, 540i
Pollinator
defined, 530
examples, 536–537, 537i, 545i
floral structure and, 536–539
Pollution
air, 531. *See also* Air pollution
Polysaccharide
examples, 502, 508, 524, 530
Pome, 542t
Poplar, 512
Poppy, 514, 514i, 542t, 550i
Population
farmland erosion, 525, 525i
Populus, 512i
P. tremuloides, 546, 547i

Porcupine, 501i
Positive feedback mechanism
defined, 495
nature of, 495
Potassium
plant nutrient, 524t
Potassium ion (K+)
ion exchanges in soil and, 524
plant function and, 524, 524t
stomatal action and, 531, 531i
Potato, 546t
Predator avoidance
spines in, 501, 501i
Predator–prey interaction
Cobra lily/insects, 523i
Venus flytrap/fly, 522, 522i
Pregnancy
miscarriages, 555
Pressure
in phloem, 532–533, 533i
translocation and, 532–533, 533i
transpiration, 523, 528–529, 529i, 535
turgor, 530–531, 531i
Pressure flow theory, 532–533
Primary growth, plant
defined, 505, 507
root, 507i, 514–515, 514i
shoot, 507–513, 507i, 510i, 516, 516i
Primary phloem, 510–511, 510i, 511i, 514–515. *See also* Phloem
Primary root, 506i, 510, 514, 514i, 515i, 552, 552i, 556
Primary shoot, 510, 552i, 553
Primary wall
formation, 554, 554i
Primary xylem, 510–511, 510i, 511i, 514i, 515. *See also* Xylem
Profile, soil, 525, 525i
Prop root, 552i, 553i
Propagation, 546–547
Protein, 524t. *See also* Transport protein
catalytic. *See* Enzyme
Protein metabolism
plants, 532
Protoderm, 507, 507i, 510i, 514
Prunus, 538, 538i, 540i
Pseudotsuga menziesii, 504, 504i, 505i, 560–561, 561i

Q

Quaking aspen, 546, 547i
Quercus, 512i, 519i

R

Radial section, plant, 507i
Ragweed pollen, 539, 539i
Ranunculus, 515i
Raspberry, 542t
Ray
in wood, 516, 516i
Ray initial, 516, 516i
Receptacle (flower), 538, 538i, 539i, 543i
Receptor, sensory
defined, 494
Red buckeye, 512i
Red oak, 519i
Redwood, 518, 519, 521
Regeneration
plant, 546, 546t
Reproductive system
flowering plant, 538, 538i
Research tool, examples
aphids as, 532
Resin, 496
Respiration
gas exchange and, 500
human, 491i
Respiratory system. *See also* Lung; Respiration

function, 491
homeostasis and, 491
human, 491i
organ interactions and, 491, 500
Rhizobium, 527i
Rhizome, 546t
Rhythmic leaf folding, 497, 497i
Rice, 546t, 547, 550
Robinia, 512i
Rock needlebush, 530i
Root
  absorption by, 526–527
  adventitious, 514
  defined, 506
  dicot vs monocot, 515, 515i
  direction of growth, 556, 556i
  embryonic source, 507, 507i
  epidermis. *See* Epidermis, plant
  functions, 506, 514–515, 526–529,
    526i, 527i, 529i
  lateral, 500i, 514, 514i, 515, 515i
  primary, 506, 507, 514i, 515i, 552,
    552i, 556
  structure, 506i, 514–515, 514i, 515i
  woody, 506, 507, 516–519, 516i,
    517i, 518i
Root cap, 514, 514i, 556, 556i
Root hair
  damage, transplants, 514–515
  defined, 514, 526
  function, 514, 514i, 526
  structure, 506, 526, 526i
Root nodule, 527, 527i, 552i
Root system
  defined, 506
  extent of, examples, 514–515, 514i
  function, 501, 506, 523
  location in plant, 490i, 506i, 529i
  mycorrhizae. *See* Mycorrhiza
  primary growth, 506, 507
  secondary growth, 506, 507
  taproot vs fibrous, 514, 514i
  transpiration and, 528–529, 529i
  water uptake by, 490–491, 490i,
    526–527, 526i
Root tip, 507, 507i, 514, 514i, 556i
Rose, 506, 509, 539i, 546t
Roundup (herbicide), 562i
Runner (aboveground stem), 546, 546i
Runoff, 525i
  clay soils and, 525
Rye plant, 515
Ryegrass, 509, 512, 514

S

Sacred lotus, 503i
Saguaro, 536i, 545, 545i
Salicylic acid, 554t
Salisbury, F., 497i
Salix, 515i
Salmon River, Idaho, 535, 535i
Sand, in soil, 524
Sap, 518
Sapwood, 518–519, 518i
Satter, R., 558
Sclereid, 508, 508i, 520
Sclerenchyma, 506, 508, 508i, 520t
Secale cereale, 561
Secondary cell wall, 508, 508i, 509, 519
Secondary growth, plant, 505–507, 505i
  accumulation, woody plants,
    516–517
  characteristics, 507, 516–519,
    516i–519i
Secondary phloem, 516–517, 516i,
    517i, 518
Secondary xylem. *See also* Wood
  embryonic source, 516–519, 516i–519i
  in roots, 516–517, 517i
  in shoots, 516–517,
    516i, 517i
  vascular cambium and, 516–518

Sedge, 544
Seed
  characteristics, 508, 509i, 541,
    542–543, 542i, 542t, 546
  defined, 543
  dicot. *See* Dicot
  digestion, animal gut, 542, 544
  dispersal, 537, 543, 544, 544i
  formation, 516, 542–543
  germination, 505, 506, 509, 514, 516,
    524t, 552, 552i, 553
  monocot. *See* Monocot
  orchid, 544
Seed–bearing plant
  characteristics, 506
Seed coat, 541i, 542i, 543, 544
  sclereids, 508
Seedless grape, 550, 550i, 551
Seedling, 541i, 550, 552, 552i, 553, 556i
Selective gene expression
  in plant development, 553–554
Senescence, 560, 560i
Sensory neuron, defined
  function, 499
Sensory pathway
  nature of, 499, 499i
Sepal, 499, 538, 538i, 543i
Sequoia
  adaptation to fires, 518i
  extent of secondary growth, 516, 518i
Sequoia sempervirens, 516, 518i
Set point, for feedback control, 494
Shell
  coconut, 508
Shepherd's purse, 542, 542i
Shoot
  defined, 506
  direction of growth, 556, 556i
  embryonic source, 507, 507i, 510, 510i
  floral, 537
  formation, 510, 510i
  primary. *See* Primary growth, plant
  secondary. *See* Secondary
    growth, plant
Shoot system, plant
  aspen clone, 547i
  defined, 506
  function, 506
  structure, 506i, 510–511, 510i, 511i
Shoot tip, 506i, 510i, 542i, 546, 547, 555
Short–day plant, 558–559, 559i
Sieve plate, 509i
Sieve tube
  characteristics, 508, 509i, 532
  defined, 532
  formation in root, 514i
  function, 532–533, 533i
  structure, 532, 532i, 533i
  in wood, 518
Sieve–tube member, 508, 509, 509i,
    511i, 532, 532i
Signal reception, transduction
  response. *See also* Cell
    communication
  nature of, 498–499, 554
Silt, 524, 535
Simple fruit, 542t, 543
Simple leaf, 512, 520
Simple tissue, plant
  categories, 508, 520t
  characteristics, 506, 508, 508i, 509i
Sink, translocation, 532–533, 533i
Sleep movements, plant, 558–559, 559i
Slime mold, 498
Smell, sense of, 536
Smith, S., 539
Smog, 531, 531i
Softwood, 519, 519i
Soil
  defined, 524
  moisture, 523
  nitrogen–deficient, 527, 527i, 560
  nutrient, 525i

nutrient losses from, 524, 524t
  plant function and, 524–527
  properties, 525
Soil horizon, 525, 525i
Soil profiles, 525, 525i
Solar tracking, 564, 564i
Solute–water balance, 501
Somatotropin (STH or GH)
  growing larger cattle with, 564
Sonchus, 533, 533i
Source vs sink, in translocation,
    532–533, 533i
Sow thistle, 533, 533i
Soybean, 560i
  growth experiment, 527i
Sperm
  flowering plant, 537, 540, 540i
Spinach, 558, 559, 559i
Spongy mesophyll, 513, 513i
Spore, plant
  megaspore, 540, 541i
  microspore, 540, 540i
Sporophyte
  defined, 538
  development, 541i, 542, 542i, 552,
    552i, 553
  examples, 540i
Stalk, leaf, 512, 512i
Stamen, 538i, 539
Starch, 524t
  breakdown by plants, 532
  storage in plants, 532
Starch grain, 556
Statolith, 556, 556i
Stem
  dicot vs monocot, 509i, 510–511, 511i
  direction of growth, 556, 556i
  formation, 510, 510i
  functions, 506, 507, 520
  hormones and, 550–551,
    554–555, 554t
  plant, 491, 506, 507, 510–511, 511i
  primary growth, 510–511, 510i, 511i,
    516, 516i
  secondary growth, 516–519, 516i–519i
Steward, F., 546–547
Stigma, 539, 540i
Stimulus, 494
Stoma (stomata), 497
  abscisic acid and, 555
  defined, 509
  function, 513, 523, 530–531,
    530i, 531i
  locations, 497, 513, 513i
  structure, 490, 509, 513i, 530, 531i
  water loss at, 528, 530–531
Stone cell (sclereid), 508i
Strawberry, 542t, 543, 543i
  vegetative reproduction, 546, 546i
Structural organization, 490–491
Style, flower, 538i, 539, 540i
Suberin, 509, 518, 560
Succession, ecological
  after volcanic eruption, 504–505, 504i
Succulent, 512, 530i
Sucrose
  transport in plant, 532–533, 533i, 534i
Sugar, simple
  transport in plants, 513, 513i
Sugar maple, 518
Sugarcane, 509
Sulfur
  plant growth and, 524t
  used as insecticide, 562
Sunflower, 508i, 542t
  gravitropic response, 556i, 564i
Sunlight
  phototropism and, 556–557, 557i
  plant adaptations to, 512–513, 517,
    531, 552, 553
Surface–to–volume ratio
  defined, 500

internal transport and, 500
  of leaves, 512
  nature of, 500
Swamp, plant growth in, 525
Symbiosis. *See also* Mutualism
  plant/bacterium, 527, 527i
  plant/fungus (mycorrhiza), 527
  protistans engaged in, 490
Syringa, 561i
Systemin, 554t

T

2,4,5–T herbicide, 555
2,4–D (2,4–dichlorophenoxyacetic
  acid), 555, 562i
Tangential section, plant, 507i
Tannin, 518
Taproot system, 514, 514i
Temperature, environmental
  flowering response and, 561, 561i
  plant development and, 552–553,
    560–561
  plant dormancy and, 560–561
Tendril, 557, 557i
Tension, defined, 528
Terminal bud, 506i, 516i
Test, observational. *See also*
  Experiment, examples
  rhythmic leaf movements, 497, 497i
  root nodule effects on growth, 527i
  tropisms in plants, 556i, 557i
Theobroma cacao, 544, 544i
Thigmotropism, 557, 563
Tiplady, B., 496
Tissue
  animal. *See* Animal tissue
  culture propagation, 537,
    546–547, 546t
  defined, 490
  plant. *See* Plant tissue
Tissue, plant
  complex, 505–509, 506i–509i
  dermal, 505, 506, 506i
  formation in roots, 514–515
  formation in shoots, 510, 510i
  ground, 505–506, 506i
  simple, 505–506, 508, 508i
  vascular, 505–506, 506i
Tissue culture propagation, 537
  orchids, 546t, 547, 547i
  Steward's experiments, 546–547
Tomato plant
  body plan, 490i, 506i
  fruit, 542t, 543, 555
  salt–tolerant, 492, 492i
  seedling, phototropism, 557, 557i
Topsoil, 525, 525i
Toxin
  defined, 562
  plant, 512
  synthetic, 562
Trace element, 524
Tracheid, 520t
  formation, 554i
  function, 508, 528, 528i
  in secondary xylem, 516i, 519
  structure, 508, 509i, 528, 528i
Transduction, 498–499, 499i
Translocation, 523
  defined, 532
  distribution pathway, 533i
  pressure flow theory and, 532, 533i
Transpiration, plant, 523, 528–529,
    529i, 535
Transport, internal, 500
Transport line, 500, 501i
Transport protein
  plant, 526, 526i
  vascular cylinder, 526, 526i
Transverse section, plant, 507i, 508i
Tree
  compartmentalization, 496, 496i

Tree farm, 519
Tree ring, 519, 519i, 521
Trifolium, 512i
Tropical forest
  identifying ecoregions for,
    493, 493i
Tropics
  growth rings and, 519
Tropism, 556–557, 556i
Tuber, 546t
Tulip, 500, 500i, 546t, 558
Turgor pressure
  defined, 530
  guard cell function and, 530–531, 531i
  in plant cell growth, 554
  plant wilting and, 530
Turner, T., 500, 500i

U

Ultraviolet radiation
  effects on organisms, 488
  floral patterns and, 537, 537i
  sensory receptors for, 537
Ursus maritimus, 492i

V

Vaccine, 506
Vascular bundle, defined, 510
  monocot vs dicot, 509i, 510–511, 511i
Vascular cambium
  activity at, 507–519, 516i, 517i
  bark relative to, 518–519, 519i
Vascular cylinder
  formation, 515i, 518i
  function, 514–515, 526, 526i, 529i

Vascular plant, defined
  characteristics, 506–517, 506i, 517i,
    522–533, 526i–533i, 538–546, 538i,
    544i, 552–561, 552i, 556i–559i
  classification. *See also* Appendix I
Vascular ray, 516–517, 516i, 517i
Vascular tissue. *See also* Phloem;
    Xylem
  dicot vs monocot, 509, 509i
  early vs late wood, 519, 519i
  floral, 538
  functions, 500, 506, 506i, 507, 538
  hardwood vs softwood, 519, 519i
  meristematic sources, 507, 507i
  primary, 507, 507i
  root, 506, 506i, 514i, 515, 515i
  secondary, 507, 507i, 516–519
    517i, 519i
  stem, 506, 506i, 508, 510–511, 511i
Vegetative growth, 537, 546
Vegetative propagation, 546–547, 546t
Vein
  human, 500, 501i
  leaf, 501i, 512i, 513, 529i,
    533, 533i
Venus flytrap, 522–523, 522i
Vernalization, 561
Vessel, xylem, 508i
  characteristics, 508
  formation in root, 514i
  function, 508
  in stem vascular bundle, 511i
  transpiration and, 528, 528i, 529i
Vessel member, 508, 509i, 528, 528i
Vine, 551
Virginia, Tidewater region,
    521, 521i

Vitis, 550
Volcano
  eruption, 504–505, 504i

W

Wall cress, 498, 498i, 564
Walnut, 516i
Water. *See also* Water molecule
  in osmosis. *See* Osmosis
  properties, 528, 528i, 529i, 535
Water conservation
  plant, 518–519, 530–531, 530i, 531i
Water lily, 500, 500i
Water molecule
  hydrogen bonding by, 528, 529i
Water pollution
  frog deformities and, 562i
Water potential, 531
Water–solute balance
  plants, 524t
Water transport, plants
  cohesion–tension in, 528–529, 529i
  compartmentalization and, 519
  nature of, 508–509
  tracheids and, 528, 528i
  vessel members and, 528, 528i
Watermelon, 542t
Watershed
  Mississippi River, 525
Wavelength
  in photoperiodism, 558, 558i
Wax, 509, 544
  in Casparian strip, 526, 526i
  in plant cell wall, 530, 530i
Weigel, D., 564

Went, F., 556
Wheat, 509, 542, 546t, 547
  domestication, 546t
White–winged dove, 545
Willow, 504, 515i, 539
Wilting, 524t, 530
Wood
  classification, 518–519
  early vs late, 519
  hardwood vs softwood, 519, 519i
  heartwood vs sapwood, 518, 519i
  meristematic sources, 507, 516–517
  products, 519, 519i
  uses, 519i
Woody plant, 516–517, 516i, 517i

X

Xylem, 500, 506
  components, 508–509, 508i, 528, 532
  defined, 508
  function, 508, 509i, 520t, 523, 528
  secondary, 516, 517, 517i, 518, 556
  structure, 506, 507i, 514–515, 515i,
    516–519, 516i–519i
  in transpiration, 528–529, 529i

Y

Yellow bush lupine, 497, 497i

Z

Zinc, 524t
Zucchini, 535
Zygote
  flowering plant, 541–542, 542i

Index of Applications

A

Agent Orange, 555
Agriculture
  erosion and, 525
  herbicides, 562, 562i
  pesticides, 562, 562i
  pine farms, 519i
  plant hormones and, 550–551,
    550i, 555, 555i
  salinization and, 492, 492i
  soils and, 525–535, 525i
  vernalization in, 561
Air pollution
  stomata and, 531, 531i
Alachlor, 562i
Alar, 562i
Allergic rhinitis, 539
Allergies, 539
Asthma, 562
Atrazine, 562i
Autoimmune responses
  multiple sclerosis, 499

C

Cattle grazing
  erosion and, 525
Chlordane, 562i
Chrysanthemums, growing
  commercially, 559
Cloning
  plant, 546-547, 547i

Crop production
  commercial tomatoes, 492, 492i
  dispersals, crop species, 544
  dormancy requirements, 561
  fertilizers (*See* Fertilizers)
  pesticides and, 562, 562i
  plant hormones and, 550, 551
  salt tolerance and, 492, 562
  toxic metals and, 562

D

2,4-D, 555, 562i
DDT, 562, 562i
Dendroclimatology, 521, 521i
Drought
  Virginia colonists and, 521, 521i

E

Erosion, 525

F

Fertilizers, 525, 525i
Freeway landscape plants, bad
  choices, 559
Frog deformities, drastic changes
  in environment and, 562i
Fungicides, 562

G

Genetic engineering
  crop plants, 562
Growth hormone in cattle feed, 564

H

Hardwood, uses of, 519i
Hay fever, 539
Hemoglobin
  HbS, 534, 534i
Herbicides, 562, 562i
Hives (allergic reaction), 562
Human body
  respiratory system, 490i
Human history
  Roanoke Island colonists, 521
Human population
  food sources for, 562

I

Infections, fungal agents
  Armillaria, 496, 496i
Insecticides, 562, 562i
Insects, as food competitors, 562

L

Lasso (herbicide), 562i
Leukemias, 555
Lost colony, Roanoke Island, 521, 521i
Luna, old-growth redwood, 521

M

Malathion, 562i
Miscarriage, 555
Mt. Saint Helens eruption, 504, 504i
Mountain climbing, 488, 488i
Multiple sclerosis, 499

O

Orchids, commercial propagation
  method, 547
Organophosphates, 562i

P

Pesticides
  crops and, 562, 562i
  effects on humans, 562
Pine farming, 519i
Pregnancy
  miscarriages, 555

R

Roundup (herbicide), 562i

S

Salinization, 492, 492i
Softwood, uses of, 519i

T

Turner, Tina, 500, 500i

U

Ultraviolet radiation
  harmful effects, 488

W

Water contamination
  frog deformities and, 562i
Wood products, 519, 519i